A GIFT OF GEOLOGY

A GIFT OF GEOLOGY

ANCIENT EGYPTIAN LANDSCAPES AND MONUMENTS

Colin D. Reader

The American University in Cairo Press
Cairo · New York

First published in 2022 by
The American University in Cairo Press
113 Sharia Kasr el Aini, Cairo, Egypt
One Rockefeller Plaza, 10th Floor, New York, NY 10020
www.aucpress.com

ISBN 978 1 649 03218 8

Library of Congress Cataloging-in-Publication Data

Names: Reader, Colin D., author.
Title: A gift of geology : ancient Egyptian landscapes and monuments / Colin D. Reader.
Identifiers: LCCN 2022017572 | ISBN 9781649032188 (trade paperback) | ISBN 9781649032195 (pdf)
Subjects: LCSH: Geology--Egypt. | Geology--Egypt--History. | Landscapes--Egypt. | Building,
Stone--Egypt. | Quarries and quarrying--Egypt--History. | Egypt--Civilization--To 332 B.C.
Classification: LCC QE328 .R33 2022 | DDC 556.2--dc23/eng20220722

1 2 3 4 5 26 25 24 23 22

Designed by Sally Boylan
Printed in China

CONTENTS

IMAGES

Figures

Plates

INTRODUCTION

I had expected my first night in the desert to be cold but had not been prepared for just how cold it would be. As the first rays of the early morning sun filtered through the gently wafting fabric of the tent, I began to think about the expedition into Egypt's Eastern Desert that lay ahead of me. This was my first taste of desert travel and, although I had already visited the Nile Valley in Egypt several times, I had no real appreciation for what a trek through the country's deserts would involve. As I welcomed the warmth of the morning sunlight, I tried to imagine the journey before me, but had little concept of the wonders I was about to experience.

My interest in ancient Egypt had been steadily growing. Initially, I had focused on the obvious: the Great Sphinx of Giza and the incredible pyramids. For me, a geologist working in the construction industry, there were many mysteries to solve: How had a largely agricultural community over 4,000 years ago managed to quarry and place millions of stone blocks to form the geometric perfection of the pyramids, and perhaps more importantly, why? And how had the same people developed skills working with gold and other metals that still manage to enthrall an increasingly skeptical modern world? Despite my best endeavors over many years, I am still a

Figure i. First night in the desert.

long way from having the answers to questions such as these, but maybe I caught a glimpse of their motivation in that first desert sunrise, as the sun's rays transformed the world around me from a cold, dark, hostile place into a world of warmth and light and opportunities. Combined with the life-giving waters of the great River Nile, it was surely the unfaltering daily cycle of the sun that inspired this magnificent ancient culture to flourish in such a spectacular way.

The sun god Re was foremost among the complex pantheon of ancient Egyptian deities, and most of the great monuments the pharaonic civilization left behind appear to have been built to pay homage to Re, or to the great god's representative on Earth, the pharaoh. It is evident that the ancient inhabitants of the Nile Valley recognized the debt that they owed to Re. The individuals who worked on the construction of the pyramids were not slaves as Hollywood would have us believe, and it is clear from the records they left behind that they were immensely proud of their work: quarrying huge volumes of limestone to build these enormous, complex, and mysterious structures in order to honor both their great king and their sun god. While vast teams labored on the construction of Egypt's great pyramids and temples, expeditions were dispatched into the arid mountains far to the east of the Nile, to quarry for gold and other natural treasures. Because gold is the color of the sun's rays (plate 1) and will never tarnish, the pharaonic Egyptians regarded it as the fiery flesh of Re. Although undertaken in the name of the pharaoh, these quarrying expeditions were ultimately intended to glorify this great god of the sun.

This is not primarily an Egyptology book, although inevitably we will explore many of the ancient monuments of this amazing land. What I have attempted in the chapters that follow is to weave together our current understanding of the separate disciplines of Egyptian geology and Egyptology to illustrate how the civilization of pharaonic Egypt benefited from the landscape in which it developed. My challenge, perhaps, is to convince you that without the many accidents of geology that shaped the Egyptian landscape, the story of this great civilization would have been very different and, if I am right, far less enthralling. This book has been written for a generally nontechnical readership, those of you who have wondered about the natural environment through which the River Nile flows, or are curious about the role the environment has played in nurturing the great civilization of the Nile Valley. I also hope that students and academics will appreciate the summaries of a number of complex issues, such as the evolution of the River Nile and the inclusion of recent research, particularly for the desert regions of Egypt.

While preparing the book, I drew heavily on my personal experiences in Egypt, from initial visits as a deeply interested tourist to the later, more specialist trips, including several years as a project geologist, carefully mapping the soils and rocks at Saqqara. My hope is that this personal focus will enrich the story as it unfolds, but inevitably, it means that less attention can be given to important sites that I have not yet visited or to subjects for which I have little or no direct experience. For those wanting to know more about a particular subject, a comprehensive "Further Reading" list is included with appropriate references provided in the main text.

The book opens by exploring the basic principles of geology, particularly those processes that have influenced the development of the Egyptian landscape. We will then explore how that landscape has evolved over geological time, stopping to examine the very different environments that have existed in Egypt at key points in the remote past, and the evidence for the animals that flourished in those environments. We will also examine the Earth-shattering upheavals that, in more recent geological periods, have shaped the landscape, ultimately leading to the development of the great River Nile. Beginning in chapter 6, we will explore Egypt's modern landscape, slowly introducing the activities of humankind into the story of Egypt's past. Later chapters will focus on issues that are key to the geological themes of this book: quarrying and the early use of stone in building. In the penultimate chapter, we will examine the possible influence of landscape on the development of important early pharaonic sites, focusing on the vast necropolis at Saqqara but also considering the potential implications of these ideas for later pharaonic monuments, theology, and tradition.

Trying to integrate our knowledge of Egyptology and geology has been a challenge, particularly because these two disciplines operate over very different timescales. Archaeologists and Egyptologists focus on the last several thousand years, while geologists focus on events that have played out over many millions of years. Given the immense backdrop of geological time, it has sometimes proved difficult to accurately place even significant archaeological events. I hope I haven't been too heavy-handed with the facts when trying to combine these very different timelines, or when trying to weave together different avenues of research. Another great and ever-present challenge has been that ongoing fieldwork often leads to new information. In the course of writing this book, a number of spectacular new archaeological finds have been announced in Egypt, and the possibility exists that once these or other new discoveries have been more comprehensively studied, our current views and understanding may need to be revised.

When I first began to take an interest in ancient Egypt, I was surprised that so many people had questions about the geology and landscape of the country. Many of the Egyptology societies that flourish in the United Kingdom and elsewhere have invited me to speak on the subject, and I want to thank all the warm and generous individuals I have met, particularly those who have raised difficult issues that have spurred me on to new channels of research. Special thanks go to Bob Partridge, who, although no longer with us, was a great inspiration for me and also for the many members of the Manchester Ancient Egypt Society, of which Bob was cofounder and long-time guiding light. I would also like to thank Ancient World Tours (www.ancient.co.uk) and the incredibly knowledgeable people who travel with them, without whom many of the experiences I share in this book would not have been possible. I must also acknowledge the opportunities presented to me by the late Ian Mathieson and the Saqqara Geophysical Survey Project, who allowed me to join them as their project geologist and to see firsthand what is required to undertake survey and excavation work in Egypt. My involvement with the SGSP gave me an insight into the ancient world that I simply could not have gained in any other way.

Although we have enjoyed holidays to Cairo and Luxor, I would also love to share my desert experiences with my family, but security concerns prevent desert travel at the moment and I cannot see the situation changing in the near future. This book is therefore dedicated to my family, who have generously allowed me so much time to pursue my interests in Egypt. Huge hugs for being so patient and understanding.

Unless stated otherwise in the image credits at the end of the book, all the illustrations are my own, as are the views and opinions that have been expressed.

CHAPTER 1
THE EGYPTIAN LANDSCAPE

Modern Egypt's strong cultural and historical ties with the Middle East make it easy to overlook the fact that Egypt is part of Africa. Including the Sinai Peninsula, the country occupies about one million square kilometers of the northeast of Africa, with coasts along the Red Sea and the Mediterranean, and land borders with Libya, Sudan, and Israel (fig. 1.1). The vast majority of Egypt's landmass is desert, with only about 6 percent of the country occupied by water (primarily the River Nile), settlements, and agricultural land. In many respects, today's landscape would have been familiar to the ancient Egyptians, with rich agricultural lands along the banks of the river (fig. 1.2) contrasting with the barren expanse of the surrounding deserts (fig. 1.4). The greatest change to the landscape over the last few thousand years has been the development of urban areas, particularly Cairo, a city of wonderful contrasts with its medieval mosques and modern skyscrapers.

Egypt: The Two Lands

It has been established from ancient texts that the ancient Egyptian worldview was one of markedly contrasting landscapes, with the fertile areas of Egypt, including the Nile Valley and Nile Delta, referred to as *kemet*, and the unwelcoming expanses of desert known as *deshret*.

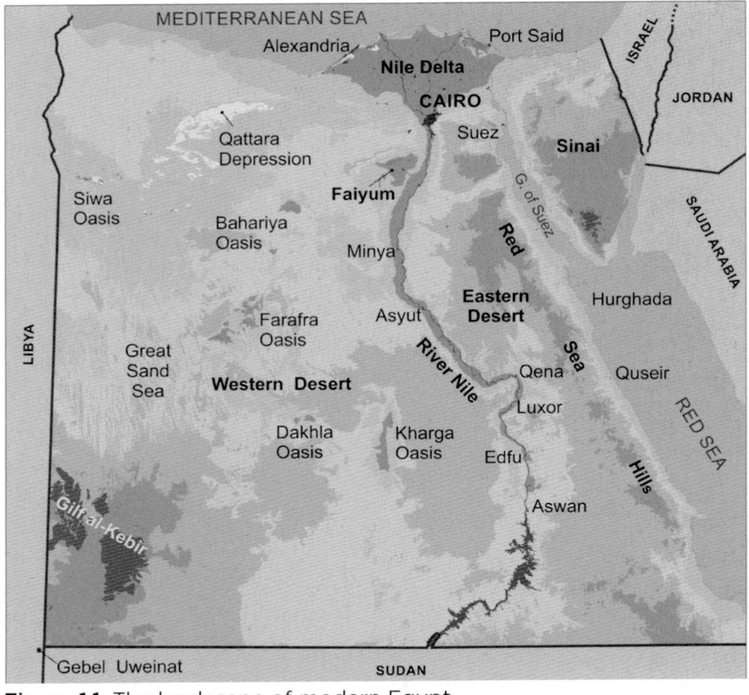

Figure 1.1. The landscape of modern Egypt.

Kemet

The ancient term *kemet*, or Black Land, referred to the agriculturally rich areas of Egypt: principally the narrow and elongated Nile Valley and the triangular expanse of the Nile Delta in the north. To these we can add the fertile areas of the Western Desert oases and of Faiyum (fig. 1.1). *Kemet* was the crucible in which the civilization of ancient Egypt developed and flourished and was regarded by the ancient Egyptians as an orderly and very benevolent place to live. In the Nile Valley and Nile Delta, the fertility of *kemet* was directly linked to the River Nile, and although the river rises far to the south of Egypt, for the ancient Egyptians the source of the Nile was at the First Cataract in Aswan (fig. 1.1). Here, the river is dominated by a wide outcrop of resistant granite rocks (fig. 1.3) that have forced the river to divide into a series of narrow streams. In places these

Figure 1.2. Rich agricultural lands bordering the River Nile.

streams are so shallow, the river cannot be navigated. The distinctive landscape at Aswan, and its ancient identification as the source of the Nile, afforded the area huge religious importance from the very earliest stages of the pharaonic era. The First Cataract at Aswan is the most northerly of a series of similarly constrained sections of the river that contrast markedly with the landscape along the rest of the Nile Valley through Egypt. For nearly 900 kilometers, from Aswan to the base of the Nile Delta at Cairo, the great river flows along a single broad and navigable channel that, unusually for a major river, has no tributaries (see chapter 6).

Except for some stretches, such as the prominent Qena Bend near Luxor (fig. 1.1), between Aswan and the Nile Delta the great river runs generally northward between towering cliffs, with barren desert beyond. Along large sections of the Nile Valley, these cliffs are set back many kilometers from the river, yet in some places the river and the cliffs meet with dramatic effect, with the creamy white limestone towering over the deep blue waters of the great river. We will look at the processes that led to the formation of the Nile cliffs in chapter 5, but it is clear that the ancient Egyptians appreciated their dramatic beauty, choosing sites such as Deir al-Bahari (plate 2) for the construction of tombs and funerary temples, including the famous temple of the female pharaoh Hatshepsut.

Figure 1.3. The River Nile as it crosses the granite landscape of Aswan.

A green ribbon of fertile land—the inundation—lies between the river and the cliffs and extends along the full length of the Nile Valley. Before the construction of modern dams at Aswan, the River Nile underwent a great annual cycle that, for the people of ancient Egypt, represented one of the most fundamental elements of their world. We will explore the causes of these events in chapter 6; however, each year in late summer, floodwaters flowed north along the river, overtopping the channel and depositing a layer of rich silt across the floodplain. These silts were highly fertile and provided the inhabitants of *kemet* with an immense agricultural resource. The annual Nile flood played such a fundamental role in the lives of the ancient Egyptians that they worshiped it in the form of the gods Hapi and Satet, and the ancient Egyptian calendar was built around the three seasons of *flood*, *growth*, and *harvest*.

From the earliest stages of the pharaonic period, the inhabitants of *kemet* mastered techniques to control the floodwaters and extract the greatest possible benefit from the fertile silt that was carried with them. Depictions dating back before the time of the first pharaohs show the king undertaking symbolic duties associated with the construction of dams and levees that had been built in preparation for the annual flood. These ancient images also depict the king cutting canals to distribute the floodwaters as widely as possible across the agricultural areas of the Nile Valley. The flat-lying landscape of today's rural Egypt must look remarkably similar to those ancient landscapes, with its lush green fields, grazing cattle, and stands of palm trees (fig. 1.2). The greatest difference is that the damming of the river at Aswan in the twentieth century has allowed the annual flood to be controlled and the floodplain is no longer underwater for months at a time.

Located at the point where the Nile Valley meets the southern apex of the Nile Delta is Cairo, Africa's most populous city and the modern successor to Egypt's early pharaonic capital of Memphis. Rushdi Said has estimated that prior to the damming of the Nile at Aswan, nearly 60 million tons of sediment passed Cairo each year, carried by the waters of the Nile as it flowed north toward the sea (see Further Reading). Before discharging into the relatively calm Mediterranean, however, the flow of the river has to reduce, and so the river divided naturally into a series of branches. At the time of the Greek historian Herodotus (484–425 BC), the Delta had six or seven branches; however, many of these have silted up to leave only two today. The reduced rate of flow along the branches of the Nile Delta led to the deposition of a large proportion of the sediment carried by the river, and over tens of thousands of years, this accumulating sediment has built up to form the flat-lying fertile wedge of the Delta (fig. 1.1). The sediments of the Nile Delta are estimated to be up to 3.5 kilometers thick and straddle two geological regions, separated by a zone of faulting. The bedrock beneath the northern section of the Delta belongs to the geologically ancient European landmass, with the bedrock beneath the southern section of the Delta forming part of Africa. The enormous weight of the accumulated Delta sediments pushing down on the sea floor has led to movement along this fault zone, resulting in parts of Egypt's Mediterranean coast sinking into the sea. This subsidence continued until surprisingly recently, as can be seen at sites such as Alexandria, where underwater archaeology has revealed the remains of ancient statues, buildings, and temples dating from the later parts of the pharaonic era.

The remaining fertile areas in Egypt are the leaf-shaped Faiyum basin and the oases (fig. 1.1), which lie in a series of depressions that are strung out across the Western Desert. As we will discuss in chapter 9, the region of Faiyum is occupied by a large freshwater lake (Birkat Qarun), the origins of which lie in the geological past, when a succession of unusually high Nile floods are thought to have broken out of the Nile Valley and flooded the depression. Lake levels have varied considerably since the initial inundation of the Faiyum basin; however, engineering works undertaken during the pharaonic era brought lake levels under some degree of control, and the Faiyum basin remains an important agricultural area today. The oases have not been influenced by the Nile: it is the availability of groundwater across the floors of these depressions that has transformed them into agriculturally rich areas that have been important since the earliest parts of the pharaonic era (chapter 8).

Deshret

Recently, there has been a great deal of interest in Egypt's deserts, with scholars seeking to understand the role that *deshret*, the Red Land, played in the development of ancient Egypt. These investigations have started to reveal a great deal of previously unsuspected detail about life in the desert during the pharaonic era. We will explore many of the findings of this research in chapters 7 to 9, but it is appropriate here to discuss what this recent work tells us about the landscape of Egypt during the time of the pharaohs. The expanses of the Eastern and Western Deserts (fig. 1.4) were regarded by the ancient Egyptians as a place of chaos: the realm of the dead, broken to the west of the Nile only by the Faiyum and the oases. For these reasons, it is generally believed that the ancient Egyptians shunned *deshret*, and although they clearly went to great lengths to defend their territory along the Nile, it is questionable whether they had much interest in defining the limits of their kingdom in the deserts. Even today, defining borders in the Western Desert is difficult, as there are very few physical features that can be used. In the furthest southwest corner of the country, the border between Egypt, Libya, and Sudan is marked by the upland area of Gebel Uweinat (fig. 1.1). To the north of the geologically ancient rocks of

Figure 1.4. The arid wastes of the Western Desert.

Uweinat is another highland area, the Gilf al-Kebir. The modern border with Libya runs through the Gilf and continues northward before becoming lost in a vast expanse of drifting sand and towering dunes known as the Great Sand Sea.

It is becoming increasingly evident that although perhaps dreaded by the ancient Egyptians, the deserts were not completely ignored. Clearly, communication links were needed between the Nile Valley and the settlements in the oases; however, as will be discussed in chapter 9, evidence is emerging for a network of desert roads that extended west, far beyond the oases, and may have been used for trading expeditions from the earliest parts of the pharaonic era. It is also evident that from the earliest dynasties, mining and quarrying expeditions were sent deep into the Eastern Desert. The Eastern Desert and Sinai (chapter 7) are dominated by the Red Sea Hills (fig. 1.5), a range of jagged, high-peaked mountains that run parallel with the Red Sea coast. As we will explore in chapter 10, these mountains are rich with metals including gold, gemstones, and other natural resources that formed the basis for much of ancient Egypt's famed wealth.

Figure 1.5. The dramatic Red Sea Hills, viewed from the Eastern Desert.

Upper and Lower Egypt

The pharaonic Egyptians applied the term "the Two Lands" to Egypt from the very earliest times. Lower Egypt (confusingly, in the north) was the name for the low-lying wide expanses of the Nile Delta, with the narrower Nile Valley to the south referred to as Upper Egypt. Outside of Upper and Lower Egypt were the relatively uncharted wastes of *deshret*. The terms Upper and Lower Egypt are thought to reflect the political origins of the ancient Egyptian state. Before being united under the rule of a single pharaoh, sometime late in the fourth millennium BC, Egypt appears to have been governed as two independent states. Lower Egypt had its ancient capital at Memphis just south of modern Cairo, and Upper Egypt had its capital at Hierakonpolis (known to the ancient Egyptians as Nekhen) north of modern Luxor (fig. 1.1). Although Egyptology has provided us with a great deal of information, the surviving records from the earliest periods of ancient Egypt are sparse, difficult to interpret, and often contradictory. However, it

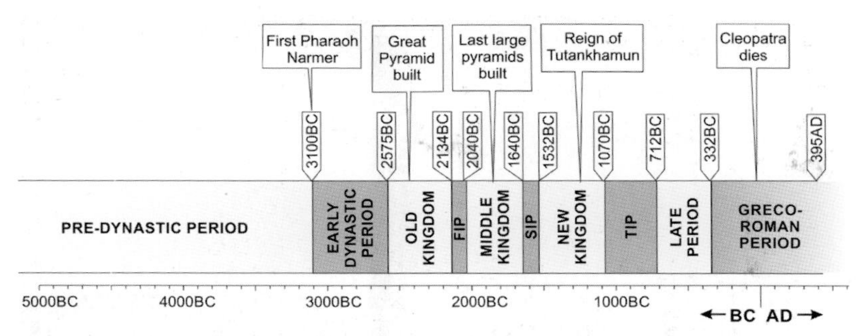

Figure 1.6. Archaeological timeline of ancient Egypt, with some key events. FIP = First Intermediate Period, SIP = Second Intermediate Period; TIP = Third Intermediate Period.

seems likely that the independent cultures of Upper and Lower Egypt had themselves developed from older city- or region-states.

It has been a long-held view within Egyptology that the unification of Upper and Lower Egypt was achieved by conflict. According to these theories, Narmer, the last predynastic king of Upper Egypt, conquered the Nile Delta to unify the two kingdoms into the world's first nation-state sometime around 3100 BC (fig. 1.6). A number of other theories suggest that the formation of the early pharaonic culture may have been influenced by foreign elements, possibly including invasion from overseas. Recent research, however, has focused on more peaceful processes of unification, possibly the result of increasingly important trading activity, which would have developed initially in the main settlements before spreading along the Nile Valley (see chapter 6). An additional theory has emerged in recent years which suggests that some elements of pharaonic Egypt may have been influenced by the cultures of the Western Desert. We will explore these ideas in more detail in chapter 9.

A Gift of Geology

As we will explore throughout this book, it was the natural benevolence of the annual Nile flood that fostered the development of the pharaonic civilization in the Nile Valley and Nile Delta. This led Herodotus to describe Egypt as a "Gift of the Nile," and in doing so, he made a very important point: that the civilization of ancient

Egypt could not have developed without the life-giving waters of this great river. It is my view, however, that the origins of the ancient Egyptian civilization need to be explored at an even deeper, more fundamental level than this. The Nile inundation was clearly critical for the development of agriculture in ancient Egypt, but we cannot overlook the fact that the particular and often uniquely benevolent features of the Nile were a direct consequence of the landscape through which the river flowed. Furthermore, we should not overlook the fact that pharaonic Egypt was more than just a rich agricultural community. As the pyramids and temples of Egypt attest, the ancient Egyptians were great engineers, developing techniques for the large-scale quarrying and working of stone to build some of the greatest and most inspiring structures mankind has ever produced. As a consequence of Egypt's landscape, the stone required to build these structures was readily available, either outcropping along the cliffs that defined the Nile Valley or, in the case of more decorative stones, lying further afield in places such as the Eastern Desert. It is my view that the ready availability of these resources fostered the development of construction in ancient Egypt, and nurtured the skills required to work even the most intractable stones into a range of artifacts from elaborate, decorative jars to statues and obelisks on a truly monumental scale. In addition, the availability of precious metals and gemstones (especially in the Red Sea Hills) allowed even the earliest phases of the pharaonic culture to develop astounding skills in jewelry manufacture, culminating in treasures that still draw huge crowds to museum collections across the world (fig. 1.7). Rather than just the River Nile, therefore, it was the courage, skill, and artistry with which the pharaonic Egyptians exploited the natural resources of the land they inhabited that shaped the development of civilization in the Nile Valley. Fundamentally, then, rather than the civilization of ancient Egypt being a gift of the Nile, we should consider that Egypt was a gift of the landscape and the geological processes that had shaped that landscape over hundreds of millions of years. It is my contention, therefore, that Egypt and the ancient Egyptian civilization should be regarded as a "gift of geology." In the chapters that follow, I hope to explain more fully why I believe this is the case.

Figure 1.7. The solid gold death mask of Tutankhamun, inlaid with precious and semiprecious stones. Most of the materials for this mask could be quarried from within Egypt, particularly the Eastern Desert and Sinai.

CHAPTER 2

AN INTRODUCTION TO GEOLOGY

Before we can start to explore the geology and landscape of Egypt, we need to introduce some of the basic ideas that underpin our understanding of the Earth and the way it has formed, from the jaw-dropping timescales involved to the enormous forces that can literally raise mountains, only to then wear them down to grains of sand.

Geological Timescales

The landscape of Egypt that was outlined in the last chapter is the product of a long and often tortured geological evolution that can be traced back about 2 billion years—a huge period of time. In archaeology, as in most walks of life, the word "ancient" generally describes events within the last 10,000 years. So, the term "ancient Egypt" nicely encompasses the 5,000 to 6,000 years from the earliest identifiable settled communities of the Predynastic Period to the death of the last pharaoh, Cleopatra VII, in 30 BC (see the archaeological timeline, fig. 1.6). Given a typical human lifespan, even archaeological timescales can be difficult to comprehend, so it is a huge challenge to try to understand geological time. To help tackle this, geologists have broken the geological history of the Earth into a series of subdivisions that are often organized graphically in a geological timescale. Figure 2.1 is an example of a geological timescale that divides the Earth's geological history as follows:

- **Eons:** The entire 4.6-billion-year geological history of the Earth is divided into two eons—the Precambrian and the Phanerozoic. This division largely reflects the view that existed when geological timescales were first developed in the 1800s, in which life on Earth was thought to have started about 600 million years ago at the beginning of the Cambrian Period (see below). The vast Precambrian Eon was therefore regarded as the time before there was any significant life on Earth. As the science of paleontology has developed, it has become increasingly clear that life started to emerge long before the end of the Precambrian; however, it has not been felt necessary to alter the basic Precambrian/Phanerozoic model of geological time to reflect this increased understanding.
- **Eras:** The Phanerozoic Eon is divided into three eras, which were selected in the 1800s to represent the three principal stages in the development of life on the planet. So, we have the Paleozoic (or "Ancient Life"), Mesozoic (or "Middle Life"), and Cenozoic (or "Recent Life") Eras. It is often easier to think of these eras as the age of arthropods and amphibians (the Paleozoic), the age of the dinosaurs (Mesozoic), and the age of mammals (Cenozoic).
- **Periods:** Each of the three eras has subsequently been divided into a series of periods, often characterized by the rock formations that were first identified at a particular location. For example, the rocks of the Cambrian Period were first mapped in Wales ("Cambria" in Latin), the rocks of the Devonian Period were first mapped in Devon, and the rocks of the Jurassic Period were first studied in the Jura mountains in France. The rocks at these and thousands of other type-localities that have been identified are vitally important for establishing the basic structure of the geological timescale. One of the greatest challenges in geology is often to establish how a particular rock layer in Egypt, for example, relates to the rocks at any given type-locality.

- **Epochs:** The more recent Tertiary and Quaternary Periods are further divided into epochs. The need for epochs in the geological timescale reflects the fact that as we get nearer to the present, we have a great deal more geological evidence to work from and can therefore look at things in much greater detail.

Although geological timescales help greatly to understand how the various eras, periods, and epochs fit together, they do not provide any indication of "scale" or the relative lengths of each geological division. You will see in figure 2.1 that the row that represents the 4 billion years of the Precambrian Eon is just as tall as

Eon	Era	Period	Epoch	Years ago	Time of day
Phanerozoic	Cenozoic	Quaternary	Holocene	10,000	
			Pleistocene	2 Million	11 59 PM
		Tertiary	Pliocene	5 Million	11 58 PM
			Miocene	24 Million	11 52 PM
			Oligocene	36 Million	11 49 PM
			Eocene	57 Million	11 42 PM
			Paleocene	65 Million	11 40 PM
	Mesozoic	Cretaceous		150 Million	11 13 PM
		Jurassic		200 Million	10 57 PM
		Triassic		240 Million	10 45 PM
	Paleozoic	Permian		290 Million	10 29 PM
		Carboniferous		365 Million	10 06 PM
		Devonian		405 Million	9 53 PM
		Silurian		440 Million	9 42 PM
		Ordovician		520 Million	9 17 PM
		Cambrian		560 Million	9 06 PM
Precambrian				4600 Million	

Figure 2.1. A geological timescale.

the row that represents the "blink-and-you'll-miss-it" 10,000 years of the Holocene Epoch. To address issues such as scale and to help understand the enormity of geological time, it can often be useful to reduce the entire geological history of the Earth to the equivalent of a single 24-hour day. This approach has been used in figure 2.2, with the clock face divided into segments, each of which represents a key geological division. The most striking result of this approach is that the 24-hour model clearly demonstrates the vast period of time represented by the Precambrian Eon, which extends from the formation of the planet all the way through to nine o'clock at night, a full twenty-one of the twenty-four hours in our model. The disadvantage of the 24-hour model, however, is that we cannot show much of the detail. For example, the period in which modern man has been present on Earth cannot be shown at the scales used; it is not possible to draw a line on figure 2.2 that is thin enough. To help combine the geological timeline and the 24-hour model, the approximate time of each geological division is given in digital characters in the

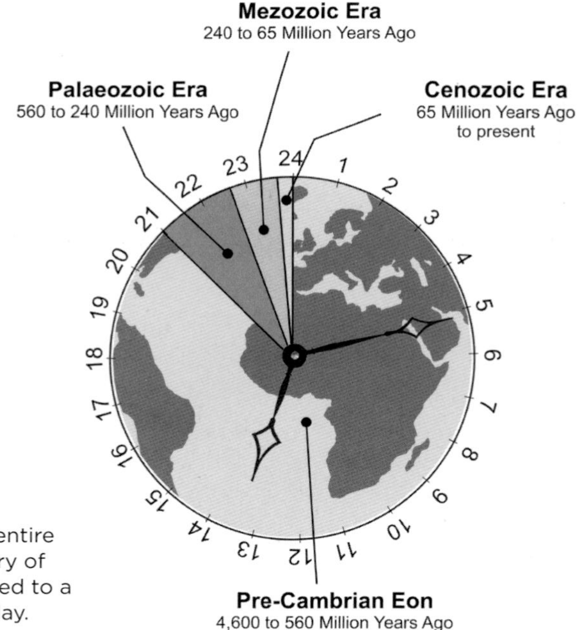

Figure 2.2. The entire geological history of the Earth, reduced to a single 24-hour day.

right-hand column of figure 2.1. Reference will also be made to the equivalent time on the 24-hour clock for a number of key geological events that are discussed in the following chapters.

Geological Maps—A Matter of Interpretation

Although this book tries to avoid being too technical, we do need a basic understanding of some of the tools that geologists use if we are to understand the role that geology has played in the development of the Egyptian landscape. Having dealt with geological timescales, we now need to look at another important tool, the geological map. A simplified geological map of Egypt is presented as plate 3, and this will be referred to frequently as our story unfolds.

Plate 3 is dominated by the solid geology of Egypt, the rock types that are exposed at the surface, and are represented by the areas of pink, gray, green, and blue on the geological map. In some areas, the solid geology is obscured by the overlying drift geology, soils that are shown on the map using various shades of yellow. Areas of drift such as the Great Sand Sea in the Western Desert, or the fertile alluvial soils of the Nile Valley and Nile Delta, are clearly visible in plate 3. No geological map is complete without a key (plate 3, bottom right), which, following the Principles of Stratigraphy (which we will explore shortly), has the oldest rocks at the bottom and the youngest rocks (and very young soils) at the top. There are also rules governing the colors used on geological maps, with colors selected to represent the basic character of the stratum that is being depicted. The following colors have been used in plate 3.

- **Yellow:** sand, alluvium, or other drift deposits.
- **Blue:** limestones.
- **Green:** sandstones.
- **Gray:** mudstones.
- **Pink:** igneous and metamorphic rocks.

There are, of course, variations to these conventions, and on more complex geological maps, different shades of the same color will be used. For example, red or purple may be used alongside pink to illustrate different igneous rock types on the same map. Although perhaps easy to ignore, the key is vitally important when it comes to reading any geological map. From the key in plate 3, we can see that the pink-colored rocks of the Red Sea Hills and the southern tip of the Sinai Peninsula are Precambrian. Without the need to refer to the geological timescale (fig. 2.1), the key to the geological map also informs us that these Precambrian rocks are the oldest strata on the map because they lie at the bottom of the key. Given their pink color, we also know that the Precambrian strata on our map are mainly igneous or metamorphic. By using the key and the geological map in this way, we can readily gather a great deal of information about the geology of an area.

The Principles of Stratigraphy

As we have already seen on both the geological timescale (fig. 2.1) and the key to the geological map (plate 3), by convention, the oldest strata are always placed at the bottom. This is a long-standing rule in geology and there is a very good reason for representing strata in this way. The majority of rocks exposed across the surface of the Earth are classified as sedimentary rocks. As the name suggests, sedimentary rocks (which include sandstones, siltstones, mudstones, and limestones) form from layers of sediment deposited under water, mainly on the sea floor (fig. 2.3, left). Over time, conditions such as the depth of the sea change, and so the nature of the sediments will also change, with coarser sediments such as sand usually laid down in shallow coastal waters (especially near river mouths) and finer silts and clays laid down in deeper offshore waters. Occasionally, perhaps rarely, animals that die and escape being eaten by predators get buried with the sediments. As these great piles of sediment accumulate over millions of years, the immense pressures that develop transform the loose sea floor deposits into a layered sequence of different rock types, referred to as a stratigraphic column (fig. 2.3, right). The process that transforms sediments into rock is known as lithification, from the Greek word *lithos*, or rock. Given that the oldest rocks are formed from the deepest sediments (the sediments that were deposited first), it follows that the oldest rocks will generally be found at the base of the stratigraphic column.

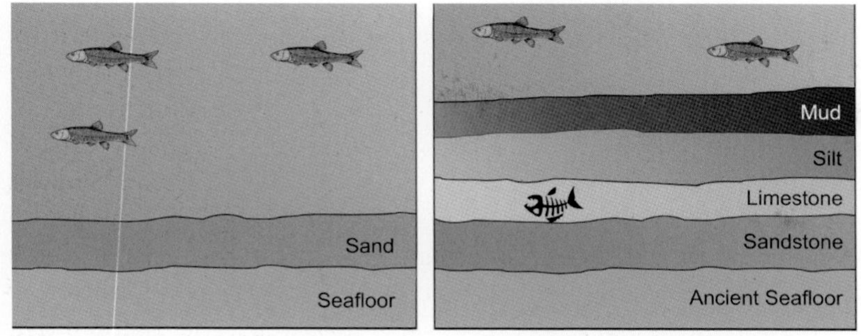

Figure 2.3. Sedimentation and the formation of the stratigraphic column.

A range of natural chemical reactions occur during lithification. In addition to influencing the development of individual rock layers, these chemical processes can also play a major part in fossilization of any plant or animal remains that have become buried (fig. 2.3, right). During fossilization, the organic material that formed the structure of the plant or animal is gradually replaced with rock-forming minerals. Although the remains of the plant or animal no longer consist of its original organic material, fossils can often preserve the original structure in incredible detail. If fossils are present, they can often be used to establish the age of a particular rock type, with primitive marine fossils such as those of the bizarre-looking trilobite being characteristic of the Paleozoic Era (fig. 2.1) and fossils from the time of the dinosaurs providing a clear indication of Mesozoic strata. Fossils can even tell us a great deal about the ecosystems in which animals and plants lived, providing additional information on the ecological conditions that existed at the relevant point in geological time.

In contrast to the dominance of sedimentation underwater, strata on land encounter mainly erosive conditions in which processes like flowing rivers or wind wear away the surface strata. Sedimentation does occur on land, for example in river valleys or in areas of sand dunes, but these conditions tend to be encountered relatively infrequently in the geological record.

Given the dominance of sedimentary rocks across the Earth's surface, most of the geological record preserves evidence for conditions that existed underwater and there is far less evidence for what was occurring on land. Despite this apparent bias in the geological record, it is possible with careful study to determine something about the conditions that existed in nearby terrestrial areas by studying marine strata. A simple example of this is the case discussed earlier, in which sandstones are often laid down in coastal waters close to the mouths of rivers. The sediments that form these sandstones have been eroded from the terrestrial rocks through which the river ran before being deposited offshore. An analysis of these sediments can tell us many things about the adjacent ancient landscape.

Other Rock Types

As well as sedimentary rocks, there are two other major rock classifications: igneous and metamorphic, both of which are important rocks in Egypt. The most widely understood type of igneous rocks are those that form when molten lava erupts from a volcano and then cools to form igneous rock such as basalt. Not all igneous rocks erupt at the surface, however. The geysers and hot springs that characterize Yellowstone National Park in the United States are powered from the heat of large active magma chambers that lie between 5 and 50 kilometers below the surface. Over geological time, activity within magma chambers such as those at Yellowstone often declines naturally and as the magma slowly cools, a solid mass of crystalline igneous rock will form at depth. The rocks that form from the cooling of molten material will take on particular characteristics, depending on the chemical composition of the magma and the rate at which cooling occurs. Given that chemistry and rates of cooling can vary considerably from one magma chamber or volcanic eruption to another, a wide variety of igneous rocks can form, many of which are subtle variations of similar types. As a result, the study of igneous rocks is a highly complex area, though perhaps not as complex as the study of metamorphic rocks.

Metamorphic rocks are formed from the alteration of other rock types, either sedimentary or igneous rocks. This alteration might arise as a result of the application of heat to the parent rock, perhaps due to the presence of a nearby magma chamber or a volcanic eruption. Alternatively, earth movements, which can range from localized earthquakes to huge deformations that occur when two landmasses collide (see below), can result in parent rocks being altered by the immense pressures involved in the collision. In many cases, metamorphic rocks result from the application of both heat and pressure. The highly complex group of metamorphic rocks is important in the study of earth processes, as some characteristic features of metamorphic rocks can be used as indicators of the history of the relevant section of the Earth's crust and the changes it has experienced. Metamorphic rocks are also very important for our study of the geology of Egypt.

A Constantly Moving Earth—An Introduction to Plate Tectonics

In 1912 a German meteorologist, Alfred Wegener, proposed the theory of continental drift. Wegener had noted (like many others before him) that the shape of the east coast of South America appeared to closely match the shape of the west coast of Africa, and that similar rock types and fossil animal groups existed on both continents. Wegener's explanation of this was that, originally, Africa and South America had been part of a single landmass and that over vast periods of geological time the continents had separated, drifting imperceptibly across the surface of the globe. Wegener could not explain the mechanisms that had driven this continental drift and geologists widely rejected his ideas. In the 1950s, however, after Wegener's death, evidence began to emerge to support the concepts he had put forward. Since then, his ideas of continental drift have been incorporated into the modern theory of plate tectonics. Unlike continental drift, in which just the continents were thought to be in motion, plate tectonics recognizes that entire sections of the Earth's crust, including the sea floor, move relative to one another. In some instances, the rates of this motion can be measured quite readily.

The mechanisms that drive plate tectonics are still not fully understood, but are likely to be connected with complex thermal processes deep inside the Earth. Many earthquakes occur when stress that builds up by the movement of one tectonic plate against another is suddenly released. However, the theory of plate tectonics does far more than simply provide an explanation for seismic events such as these. Along the middle of the Atlantic Ocean is a mountain range called the Mid-Atlantic Ridge. The Mid-Atlantic Ridge is the longest mountain chain in the world; however, its significance is easily overlooked. This is because most of its towering peaks are hidden beneath the waters of the Atlantic Ocean, with only the highest mountains standing above sea level to form islands such as the Azores and Iceland. The Mid-Atlantic Ridge has developed along a giant rift in the Earth's crust, and as molten rock is forced upward into this rift, it causes the Atlantic Ocean to widen, pushing Europe and America apart at a rate of about 2.5 centimeters a year. As the theory of plate tectonics has developed, it has been recognized that the Mid-Atlantic Ridge is one of a number of sites of sea floor spreading that are currently active around the world (fig. 2.4, top).

Given that the Earth is not becoming noticeably larger in size, it follows that in addition to areas of sea floor spreading there must be other areas in which the tectonic plates are converging. A good example of this has been identified along the west coast of South America, where a section of the Pacific Ocean (the Nazca Plate) is moving beneath the South American Plate. The diving Nazca Plate has deformed the edge of the continental plate of South America, lifting the edge of the continent to form the Andes Mountains, and as the Nazca Plate plunges beneath this coastal mountain chain, a deep offshore trench has formed in the sea floor (fig. 2.4, bottom). In simple terms, we can argue that the widening of the Atlantic Ocean at the Mid-Atlantic Ridge is being offset by the narrowing of the Pacific Ocean at places like the Nazca Trench.

Clearly, as a result of plate tectonics, the crust of the Earth is constantly moving, and it appears that this has been the case over most of the planet's geological history. The opening up of the Atlantic Ocean that was first proposed by Wegener began as recently as 120 million years ago (in the Cretaceous, about 11:20 p.m. on our 24-hour clock, fig. 2.1) and was associated with a late episode in the breakup of an earlier supercontinent known as Pangea. Pangea itself formed from the collision of two previous supercontinents, Gondwana and Laurussia (chapter 4). Although these ancient continents have come and gone, fragments of them survive as cratons, buried deeply beneath some of today's continental landmasses.

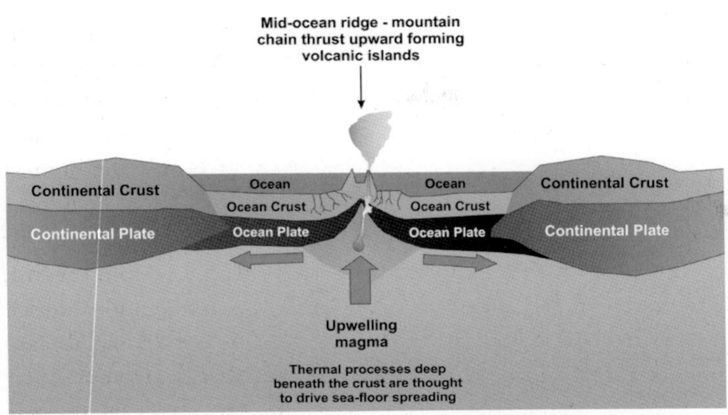

Figure 2.4. top: Sea floor spreading and the formation of mid-ocean ridges.

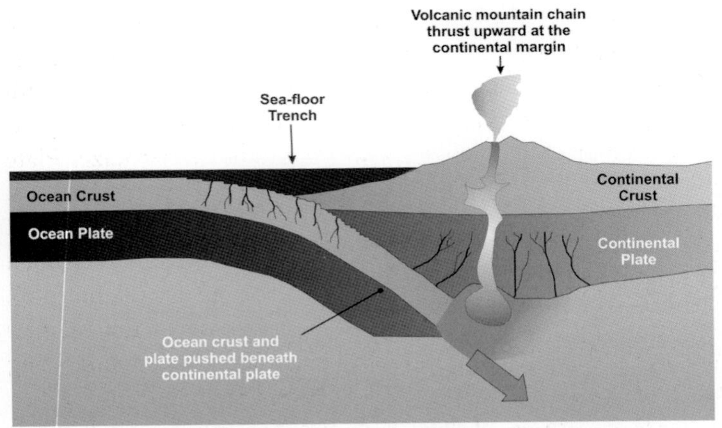

Figure 2.4. bottom: Plate convergence: the formation of deep ocean trenches and mountain ridges along continental margins.

The Development of the Egyptian Landscape

The earth-forming processes described in this chapter generally occur extremely slowly. Given the Earth's long geological history, however, there has been sufficient time for mountains to form, only for them to be eroded away and to be transported as sand along ancient rivers and deposited into ancient seas. As more time passed, these seas narrowed as continents collided and the sediments on the sea floor were thrust up into new mountain ranges; the cycle of erosion, deposition, and mountain building then started again. In order to understand the evolution of the Egyptian landscape, we need to explore these processes—to discover tales of smashed continents from times long past and of mountain building and destruction, evidence of vast floods, retreating seas, and enormous clefts in the Earth's crust that once spewed forth incredible volumes of lava, burying the landscape. As we get nearer to the present, we need to explore the evidence for enormous river systems that dwarf today's River Nile but have since disappeared beneath vast swathes of desert sand. The evidence for all these events is there to be read in the soils and rocks of Egypt, although they are often hidden and difficult to fully understand. In the chapters that follow, we will break the epic story of the evolution of the Egyptian landmass into three main acts, to reflect the Earth's great geological divisions. Naturally, we will start furthest back in time in the Precambrian Eon.

CHAPTER 3

THE GEOLOGY OF EGYPT, PART 1—THE PRECAMBRIAN

The vast period of time represented on the geological time-scale by the Precambrian Eon (over twenty-one hours of our 24-hour clock, figs. 2.1 and 2.2) ended about 560 million years ago. Of course, the Precambrian Earth looked very different than it does at present. None of today's continents had formed at this stage, and most of the landmass was distributed across the southern hemisphere, with Africa near to the south pole. Although the late Precambrian landmass shown as figure 3.1 contained most of the elements that make up modern Africa, today's African continent did not form until much later, in the Mesozoic, about 100 million years ago (or at about 11:30 p.m. on our 24-hour clock).

The surface exposures of Precambrian strata in Egypt are shown in pink on the geological map (plate 3). The oldest Precambrian rocks in Egypt are exposed in the Gebel Uweinat region in the extreme southwest of the country and are about 2 billion years old (about 1:30 p.m. on our 24-hour clock). Although slightly younger, the ancient rocks of the Red Sea Hills, which extend into the southern tip of Sinai, together with the famous granites at Aswan, also date from the Precambrian Eon. In addition to the surface outcrops of Precambrian strata shown on the geological map,

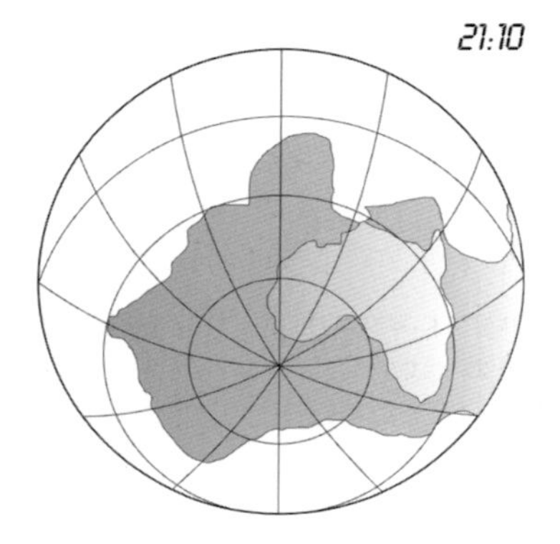

Figure 3.1. A view of the Earth's south pole as it may have looked at the end of the Precambrian, indicating the approximate position of modern Africa.

boreholes drilled deep into the Egyptian landmass have encountered Precambrian strata at depth, underlying often huge thicknesses of younger Mesozoic sandstones (shown in green on the geological map, plate 3) and Cenozoic limestones (shown in blue). These extensive Precambrian strata have played a vitally important role in the evolution of the Egyptian landmass. They are often referred to as the Precambrian Basement, a term that captures the important structural role that these ancient rocks played as younger strata were laid down on top. As we will discuss in later chapters, despite the huge gulf of time since the Precambrian, these ancient strata have had a major influence, not only on the development of the Egyptian landscape but, perhaps surprisingly, on the cultures of the Nile Valley.

Given their great age, the Precambrian strata have been exposed to a wide range of geological processes, ranging from relatively small-scale and localized earth movements to far more significant events on a global scale. With this long and tortured history in mind, it is not surprising that the Precambrian rocks of Egypt are highly complex. They include fragments of ancient continental crust, together with the remains of islands and archipelagos that were smashed against the margins of ancient continents as tectonic forces closed long-lost oceans. The ancient sea floor crust and overlying sediments that were squeezed between these colliding landmasses were raised up to form mountains that have since been eroded away. Although these ancient mountain ranges are long gone, remarkably, their remnants can still be seen in the Egyptian landscape.

Plate 4 shows a weathered outcrop of Precambrian strata, a conglomerate exposed in the central Eastern Desert (the location of the conglomerate outcrop is shown in fig. 7.1). In geology, the term "conglomerate" refers to a rock that consists of rounded gravel and smoothed cobbles embedded in a matrix of finer sand. The gravel and cobbles shown in plate 4 range in color from green-gray to orange-red, indicating that they were derived from a variety of sources. The gravel and cobbles also range greatly in size, with some of the larger pieces almost the size of a football. Originally, the components of this conglomeratic rock (the sands, gravels, and cobbles) will have been laid down as loose soils, perhaps in a river bed; however, these loose soils were subsequently lithified. If we combine our understanding of modern geological processes with a certain amount of imagination, these conglomerates can tell us much about conditions that existed in this part of Egypt in the distant Precambrian.

Rock-forming sediments generally occupy a narrow band of material sizes. For example, sandstones generally consist of sand-sized grains, mudstone consists predominantly of clay-sized sediments, and so on. Although conglomerates are not unusual, the fact that they consist of a wide range of grain sizes indicates they were laid down under a particular set of geological conditions. One example of the conditions in which conglomerates can form is when rivers flood, perhaps following a period of snowmelt. The energy of these flood waters allows coarse fragments of rock plucked from the snow-covered mountain peaks to be carried downstream, and as the fragments are tumbled and abraded, they become smoothed and rounded. Once the floods begin to subside, the more gentle flow of the river can no longer carry the larger material, and the coarse, rounded fragments fall to the sandy riverbed. Over time, these mixed deposits become lithified, turning the loose sediments into rock.

The precise age of the conglomerates in plate 4 is difficult to define, but we can assume that they were formed between 600 million and 1 billion years ago, and they provide a very graphic example of how it is possible to reconstruct the ancient environment of a specific location from just a single rock exposure. In this case, we have tentatively replaced the arid conditions of today's Eastern Desert with an alpine landscape that existed over 600 million years ago, with high snow-capped mountains and seasonally swollen rivers. But this is only part of the story. Although this alpine landscape no longer exists, we can trace its evolution even further back in time. The mountains from which the coarse gravel in our conglomerate had originally eroded were formed by great upheavals of the Earth's crust that had taken place many millions of years before our alpine rivers began to flow. The alpine landscape we have reconstructed to explain the presence of conglomerates at this location therefore leads to further questions about ancient mountains and when and how they were formed . . . taking us even further back in time.

The events that shaped the Egyptian landscape in the Precambrian are not just a matter of re-creating long-lost landscapes. There are some very real and important implications associated with developing an understanding of the distribution of some of the Earth's oldest rocks. As we discussed in chapter 2, when strata are faulted and folded in great tectonic events such as sea floor spreading (fig. 2.4, top) or plate convergence (fig. 2.4, bottom), the temperatures and pressures involved cause alterations in what may originally have been sedimentary or igneous strata, leading to the formation of metamorphic rocks. There are a wide variety of metamorphic rocks, depending on the original type of strata that have been altered and the conditions of temperature and pressure that the parent rock has been exposed to. For example, if sandstone deposits are altered by relatively low levels of heat or pressure, the quartz grains will be partially melted and recrystallize as the rocks cool or pressures normalize. This leads to the formation of the low-grade metamorphic rock quartzite, a stone that was often used by the ancient Egyptians for large statues and certain architectural elements in temples and other important buildings. Despite being a low-grade metamorphic rock, quartzite still retains elements of the parent sandstone; however, it is generally much stronger than the sedimentary rocks from which it formed. In high-grade metamorphic rocks, the entire structure of the original strata can be altered. For instance, relatively weak siltstones and mudstones can be altered, generally as a result of high pressures, to form hard, platy rocks such as schists or slates.

As well as altering the fabric of the rocks themselves, tectonic processes will also affect the nature and distribution of minerals within the rock mass. Generally, metals and other minerals are distributed fairly evenly within the Earth's crust and are thus normally present only in low concentrations. Tectonic processes, however, lead to the redistribution of these minerals, concentrating them in narrow zones such as mineral veins. A good example of this is the gold mineralization of the Red Sea Hills in Egypt and farther south into Nubia (a name that was derived from the ancient Egyptian word for gold: *nub*). This mineralization occurred shortly after the end of the Precambrian, as tectonic forces brought two ancient continents together and the sea floor between them was dragged beneath one of the continental plates, in a manner similar to that shown at the bottom of figure 2.4. As the rocks of the ancient sea floor plunged deeper, they melted and were subject to tremendous heat and pressure. Water trapped by this process became superheated, and as this superheated water was driven upward through the rock mass, metals and other minerals were removed into solution in a process called leaching. As the mineral-rich fluids moved away from the zone of plate convergence, often moving along faults and fissures in the rock (fig. 3.2), temperatures and pressures subsided, and as the fluids cooled, minerals began to crystallize. Common minerals such as quartz will have dominated these cooling fluids; however, under the right conditions, metals such as gold and copper will have been formed in association with these quartz veins.

Although I have greatly simplified the processes that led to the formation of mineral veins, these general principles can also be used to explain the formation of gemstones, a class of geological materials that were to become important for the pharaonic culture of Egypt. Gemstones such as turquoise and jasper generally form from the precipitation of mineral-rich groundwater under conditions of low temperature and pressure, with the resulting gemstone often taking the shape of the void in which it forms. If additional heat from magmatic activity is introduced into the process, crystals develop as the mineral-rich fluids cool. The chemistry of these crystals will vary enormously to produce a range of gemstones that include amethyst (a quartz mineral that derives its color from the

Figure 3.2. Mineral veins in the Eastern Desert (arrowed). Note also the small-scale faulting (dashed line) that cuts across the upper ends of these veins.

presence of iron) and beryl, which can exhibit a range of colors depending on its specific chemistry. Emeralds are a form of beryl, with the green color influenced by the presence of chromium or vanadium. In some instances, the crystals that we regard as gemstones can develop under significantly higher temperature and pressure conditions, forming directly from cooling magma rather than heated, mineral-rich groundwater. Gemstones that form in this way include topaz and tourmaline. Metamorphic processes also lead to the alteration of some rock-forming minerals into gemstones. As we have already seen, the heat from magma can alter the parent rock into which it is injected, causing the minerals in the original rock mass to recrystallize. Depending on the nature of the parent rock and the degree of this recrystallization, a range of gemstones including sapphires and rubies can develop. Under the more extreme temperatures and pressures that are encountered in tectonically active areas, particularly in areas where tectonic plates collide, minerals can develop that are unique to the prevailing conditions at that location. The contact between two tectonic plates in a small area of Tanzania, for instance, has led to the development of the blue gemstone tanzanite, which, so far, has not been identified anywhere else on the planet. All the processes described above relate to the formation of gemstones in the Earth's crust. Diamonds and peridot, however, belong to a class of rare gemstones that develop under the extreme conditions of the Earth's mantle, the component of our planet's interior that lies beneath the crust. The depth at which these gemstones form far exceeds the deepest quarries, and therefore they generally remain out of reach, only accessible at locations where deep mantle rocks have been brought to the surface, generally by volcanic activity.

The Precambrian rocks of Egypt are not just interesting because of their great age or because they contain metal ores and gemstones. It is clear from the evidence unearthed by Egyptologists over the last 150 years or so that from the very earliest flowerings of culture in Egypt, the people of the Nile Valley were venturing out into the farthest reaches of the deserts and returning with examples of the most decorative Precambrian stone. Early stone workers in Egypt developed incredible skills when working with these rocks. These skills are perhaps best illustrated by vessels in museum collections, carved from very brittle types of stone, but with walls that are so thin, the vessels glow when lit from inside. One of the main themes of this book is my belief that many of the characteristic elements of the pharaonic civilization were influenced by the ready availability across Egypt of a variety of decorative stones and other mineral resources. Before we can address these issues in more detail, however, we must consider the evolution of the Egyptian landscape since the end of the Precambrian.

CHAPTER 4
THE GEOLOGY OF EGYPT, PART 2—THE PALEOZOIC AND MESOZOIC

The Paleozoic—Long Journey North

As we saw in the last chapter, by the end of the Precambrian most of the Earth's landmass lay in the southern hemisphere (fig. 3.1), with the area of modern Cairo lying within what we would think of today as the Antarctic Circle. As we progress through the early Paleozoic, a large part of the Precambrian landmass separated, drifting north to form the supercontinent of Laurussia, with Egypt and Africa remaining as part of the southern landmass of Gondwana. From the Silurian Period onward, however (fig. 4.1), Gondwana also began to drift northward, commencing Egypt's long journey from ancient subpolar latitudes to today's position astride the Tropic of Cancer. As Gondwana approached Laurussia, there began a collision between the two supercontinents that lasted until the end of the Paleozoic, ultimately leading to the formation of the landmass of Pangea (fig. 4.2). Although it is impossible for us to fully understand the cataclysmic events that accompanied this collision, its consequences are still evident today. For example, to the west of Egypt, Morocco's snow-capped Anti-Atlas Mountains, which contain some of the highest peaks in Africa, were thrust upward during this truly Earth-shattering event.

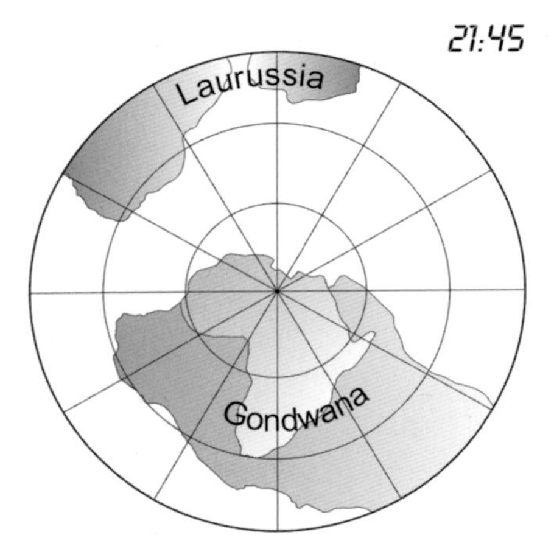

Figure 4.1. A reconstruction of the landmass around the Earth's south pole in the Silurian Period, giving an approximate distribution of the Earth's landmass.

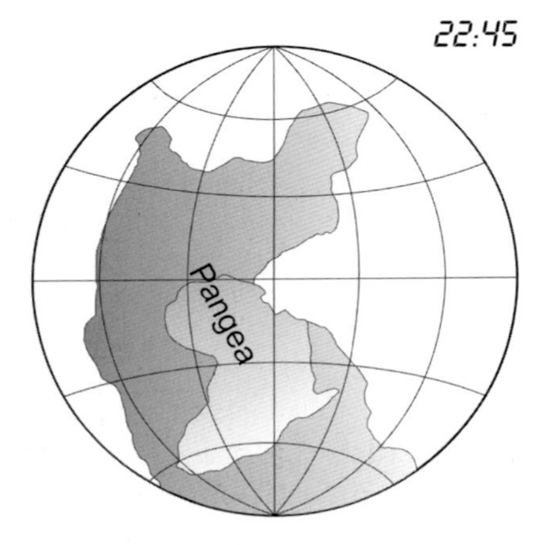

Figure 4.2. The approximate distribution of the Earth's landmass in the late Permian Period.

Although much shorter than the Precambrian Eon (fig. 2.2), the Paleozoic represents a major segment of Earth's geological history and is a tremendously significant period because of its relevance for the evolution of life on Earth. Fossils found in Paleozoic rocks from all over the world show that a wide variety of life forms emerged and flourished during this period. Some of the animals that were abundant at this time have long become extinct, while others, such as amphibians and the earliest winged insects, are largely recognizable today. There are, however, only limited exposures of Paleozoic rocks in Egypt, and those that have been identified are generally from the earlier parts of the era, up to the end of the Carboniferous Period (fig. 2.1). As the limited areas of gray on the geological map show (plate 3), the most extensive Paleozoic outcrops are in the southwest of the country, around the Gebel Uweinat and Gilf al-Kebir areas, with other limited outcrops at the northern end of the Gulf of Suez and the western part of Sinai. One explanation for the limited distribution of Paleozoic rocks in Egypt may be that throughout the era, only the areas to the west (modern Libya) and east (today's Sinai) were under water, and it was sediments deposited under these limited marine conditions that became lithified to form the Paleozoic rocks we see on the geological map. Alternatively, during the earlier part of the Paleozoic, all of Egypt may have lain under water with strata laid down across the entire Egyptian landmass. Subsequently, the central areas of Egypt may have been raised above sea level and then subjected to a prolonged period of erosion, removing the majority of the Paleozoic rocks from the geological record. Whatever the explanation, the Paleozoic is something of a "Dark Age" for Egyptian geology, with the 320-million-year gap in the geological record representing one of the biggest unknowns in the evolution of the Egyptian landscape.

The Mesozoic—The Age of the Dinosaurs

Soon after the supercontinent of Pangea had finally coalesced (fig. 4.2), it began to break apart. The rupturing of Pangea began sometime in the Jurassic Period (about 200 million years ago, or 10:57 p.m. on our 24-hour clock), and as the Earth's landmasses once more drifted apart, Egypt fell under the influence of an ever-widening ocean, known to geologists as the Tethys. The geological record indicates that during the Jurassic Period, the Tethys over Egypt was relatively shallow and covered only what were to become the northern parts of the country. By the Cretaceous Period (fig. 2.1), however, the coastline had shifted south and most of Egypt lay underwater.

Throughout this period of continuously advancing seas, a network of large rivers drained the surrounding continental regions, depositing their load of sand into the shallow coastal waters. As the position of the coast shifted southward across Egypt, a substantial thickness of sand was laid down, which became lithified to form one of the major rock units in this part of North Africa, the Nubian Sandstones. The Nubian Sandstones in Egypt are up to 3,000 meters thick and, as shown in green on the geological map (plate 3), are exposed at the surface across much of central and southern Egypt. The Nubian Sandstones are, however, more extensive than the geological map suggests, and in the northern areas of Egypt are overlain and therefore obscured by younger Cenozoic strata (shown in blue in plate 3). This sedimentary sequence can be seen clearly in the Bahariya Oasis, which is one of a number of large natural basins hollowed out of the surface of the Western Desert. The towering cliffs that form the walls of these basins generally consist of Cenozoic limestones (fig. 4.3); however, at Bahariya, the older underlying Nubian Sandstones are exposed across the floor of the oasis. This sequence is illustrated on the geological map (plate 3), with the floor of the Bahariya Oasis identified as an isolated patch of Nubian Sandstones (in green) surrounded by Cenozoic limestones (in blue).

The Nubian Sandstones are one of the world's largest known aquifers, with the porous sandstones allowing groundwater to collect in the spaces between the lithified sand grains. If it were not for the Nubian Sandstone aquifer, the Western Desert oases would be lifeless depressions, as barren as the desert that surrounds them. There is continuing debate regarding the source of the vitally important groundwater in the Nubian Sandstones. Some researchers argue that this groundwater is being replenished from areas to the south of Egypt, where the sandstone is closer to the surface and can be reached by surface waters, perhaps by seepage from lake bottoms or riverbeds. If this is the case, groundwater beneath Egypt is likely to be renewed and, providing it is not over-exploited, will remain as a valuable natural resource. By contrast, some researchers believe that the groundwater beneath Egypt originated from rains that fell during and shortly after the Mesozoic Era. This school of thought believes that the groundwater in the aquifer is not being replenished and that careful management of this finite groundwater resource will be required.

Figure 4.3. The northern cliffs of the Western Desert oases (in this case Dakhla Oasis) comprise layers of Cenozoic limestones. In this photograph the layers of limestone are obscured by banks of sand that have been blown into the oasis from the surface of the Western Desert. Older Nubian Sandstones are exposed at the base of the cliffs and across the oasis floor.

The classic terrestrial fauna of the Mesozoic are, of course, dinosaurs. The story of scientific exploration for dinosaur remains in Egypt began by accident in 1910 when a German paleontologist, Ernst Freiherr Stromer von Reichenbach, arrived in the country to begin a search for fossils of early mammals. Stromer's initial excavations in the Cenozoic strata to the northwest of Cairo and the rocks of the eastern Nile Valley near Luxor were unsuccessful, so he then turned his attention to the floor of the Bahariya Oasis, which at the time was thought to consist of exposures of Cenozoic strata. As we have already seen, however, the floor of the Bahariya Oasis consists of Mesozoic strata, and rather than fossils of small early mammals, what Stromer found was far more dramatic. In an area surrounding a small, isolated hill called Gebel al-Dist, Stromer excavated the fossilized bones of a large plant-eating sauropod (*Aegyptosaurus*—similar to a brontosaurus), and the remains of three large carnivores

(*Bahariasaurus*, *Carcharodontasaurus*, and *Spinosaurus*). Not only was it surprising to find the remains of so many animals in this relatively small area, the size of the animals Stromer found was startling. *Spinosaurus*, for example, had been a ferocious beast that was larger than *Tyrannosaurus* and, as recent studies have confirmed, appeared to hunt underwater, most probably in large rivers.

Unfortunately, the unique specimens that Stromer recovered from Bahariya were damaged on return to his laboratory in Berlin, and the outbreak of the First World War prevented him mounting further expeditions, but that was not the end of his story. In the late 1990s, Stromer's important but ultimately ill-fated work inspired a team of American paleontologists to return to Bahariya in search of his lost dinosaurs. As discussed by William Nothdurft and his colleagues, the results of this work were spectacular (see Further Reading). To add to the species that had been previously identified by Stromer, the American team identified another extremely large sauropod (*Paralititan*), the second-largest creature ever known to have walked the Earth. Paleontology has come a long way since Stromer's day, and the American team not only recovered the bones of these incredible creatures, they also examined the rocks themselves in order to reconstruct the environment in which these animals lived, 95 million years ago. Perhaps the biggest clue to reconstructing the paleoenvironment was the identification of fossilized mangroves (fig. 4.4), indicating that at the end of the Mesozoic Era, the area of Bahariya Oasis was a warm coastal environment, similar to conditions in today's Florida Everglades. *Paralititan* means "tidal giant," a direct refence to the coastal conditions identified by the American team and further confirmation that during the Mesozoic Era, the landmass that was to become Egypt was dominated by ever-changing coastlines and transgressing seas.

Throughout the spectacular evolution of the dinosaurs, the Earth's landmasses continued to drift across the surface of the planet. At this stage, as North America and South America began to drift westward, opening the Atlantic Ocean, the separate landmasses of Africa and Europe began a long, drawn-out collision that continued into the early parts of the Cenozoic. To the west, this collision would lead to the formation of the Alps in Europe and

Figure 4.4. Fossilized mangrove roots, in this case from Cenozoic strata in Wadi al-Hitan (see chapter 5).

the High Atlas Mountains in Morocco (not to be confused with the older Anti-Atlas Mountains, see above), but in Egypt the effects of the collision of Africa and Europe were less marked. As indicated in the left and right margins of plate 3, across northern Egypt this collision folded the strata of the shallow sea floor into a series of northeast-to-southwest-oriented ridges and troughs, which geologists refer to as the Unstable Shelf. The unfolded areas across southern Egypt are referred to as the Stable Shelf. As we will see in the following chapter, the development of the Unstable Shelf in the late Mesozoic had significant implications for the Cenozoic geology of Egypt and, indeed, for the archaeology of ancient Egypt. The collision with Europe may also have led to general uplift across North Africa, causing the ancient seas across northern Egypt to become increasingly shallow and the coastline to retreat northward. Toward the end of the Mesozoic, fine sediments such as silts and muds were laid down in these shallow coastal waters and were lithified to form the mudstone and shale strata of the uppermost

Cretaceous and lowest Cenozoic beds. In an area between the Farafra and Bahariya Oases, these fine sediments were dominated by calcium carbonate–rich deposits, which were laid down in warm seas that were teeming with shellfish, corals, plankton, and other sea life. Many of these life forms took calcium carbonate from the seawater to build shells and other hard body parts, and when they died, their calcium-rich remains accumulated on the sea floor. As these sediments lithified, white-to-cream-colored, chalk-like strata were formed, which have subsequently been eroded to form the dramatic and otherworldly landscape of today's White Desert (plate 5).

The extremely pure limestones of the White Desert straddle the boundary between the Mesozoic Era and the last of the three great geological divisions that shaped the Egyptian landscape, the Cenozoic.

CHAPTER 5

THE GEOLOGY OF EGYPT, PART 3—THE CENOZOIC: A LAND OF RIVERS

Although the Cenozoic Era represents only the last twenty minutes or so of our 24-hour geological day (fig. 2.2), in terms of the evolution of the Egyptian landscape, this period is perhaps the most significant. The Cenozoic rocks exposed across the surface of Egypt are shown in blue on the geological map (plate 3) and, as already discussed, were laid down in the warm shallow seas that had covered large parts of northeast Africa by the end of the Mesozoic. As the Cenozoic progressed, the sea continued to retreat northward, exposing a new landmass in its wake. Today, only limited exposures of the earliest epoch of the Cenozoic Era, the Paleocene (fig. 2.1), can be identified, often lying at the base of the limestone escarpments that overlook the Western Desert oases (see chapter 8), or in stretches along the base of the cliffs that line the Nile Valley. Perhaps the best-known Paleocene deposits are the Esna Shales, which in the area around modern Luxor are exposed in the cliffs behind Hatshepsut's mortuary temple (plate 2). The Esna Shales were also encountered during the quarrying of the royal tombs in the Valley of the Kings, where these relatively weak strata introduced a number of challenges for the ancient tomb builders (chapter 11).

One of the most significant Cenozoic epochs in Egypt was the Eocene (from about 57 million years ago [Mya], or 11:40 p.m. on our 24-hour clock). As discussed in the last chapter, in areas north of modern Asyut (fig. 1.1) the collision between Africa and Europe had folded the underlying Mesozoic strata into a series of ridges and troughs, referred to as the Unstable Shelf (plate 3). Massively thick Eocene deposits accumulated in the deeper water conditions within the troughs of the Unstable Shelf; however, the shallow water conditions over the ridges, particularly those ridges closest to the shoreline of the period, provided an ideal habitat for an unusual group of fossils that have a surprisingly important role in the archaeology of ancient Egypt. The Giza Plateau southwest of Cairo is renowned as the site of the Great Pyramid of Khufu together with other Old Kingdom pyramids, temples, tombs, and of course the Great Sphinx (plate 6). Very few visitors to Giza will be aware, however, that the existence of the plateau on which these wonders of the Old Kingdom were built owes much to marine conditions over the ridges of the Unstable Shelf, some 40 million years ago.

Remarkable Cenozoic Fossils

Geologists have identified that one of the ridges of the Unstable Shelf passes through the Giza Plateau. Forty million years ago, the shallow water conditions along the crest of this ridge were ideal for an unusual group of creatures called nummulites, the remains of which are characterized by their flattened disc-shaped shells. Although the few surviving nummulite species have shells that rarely exceed 2 millimeters in diameter, their extinct Eocene relatives were comparative giants, growing shells that were up to 160 millimeters across. The size of these shells is all the more remarkable given that the creatures that built and lived inside them were tiny, single-celled organisms that belong to a group known as protists. Another unusual feature of nummulites is that they obtained nourishment from algae that grew within the coiled shell and, in return, benefited from the protection the shell provided from predators. This symbiotic relationship was very successful, allowing nummulites to colonize in vast numbers in the shallow, warm, sunlit waters over the ridges of the Unstable Shelf. Today the shoals of nummulites at Giza survive as fossils, with the disc-shaped shells in such abundance in some areas of the Giza Plateau that the exposed limestones are almost entirely made up of nummulite remains.

In the ancient Cenozoic seas, the submarine ridge that runs through the Giza Plateau and that the nummulites exploited so successfully acted as a reef that separated deeper offshore conditions from more sheltered, inshore conditions. Sediments laid down in the deeper offshore waters developed into thick, massively bedded strata, which millions of years later during the Old Kingdom (fig. 1.6) proved to be perfectly suited to quarrying for pyramid building. In the shallow, more frequently changing conditions inshore of the reef, however, more variable beds were laid down—strata that were far less suitable for use in ancient building projects. The alternating sequence of thin sandstones and siltstones exposed by the carving of the Great Sphinx (plate 6) was laid down in one of these sheltered inshore lagoons.

Being able to interpret the strata at sites such as Giza and to recreate the paleo-environment in which the various rock types were formed is not simply an interesting exercise for geologists. Developing an understanding of the conditions that influenced the formation of specific strata exposed at the Sphinx and other ancient monuments not only allows us to understand the manner in which ancient monuments weather and erode, but can be vital when assessing the most appropriate methods of conservation.

Giza with its nummulite-rich limestones and Gebel al-Dist with its globally significant assemblage of Mesozoic megafauna remains (chapter 4) are just two of a number of important fossil sites that have been identified in Egypt. Generally, fossil remains are piecemeal, with animal carcasses broken up by predators or by currents long before they become fossilized. What makes the fossil assemblage at Wadi al-Hitan so remarkable is the exceptional preservation that has been encountered there. Wadi al-Hitan (fig. 8.1) is located about 40 kilometers southwest of the Faiyum region and has been awarded UNESCO World Heritage status. Although the site was first identified in the early 1900s, its remote location meant that it received relatively little attention until the 1980s. Since then, however, hundreds of fossils of marine creatures have been excavated, often in the position in which they died some 40 million years ago. Wadi al-Hitan is Arabic for "Valley of the Whales" and is named after the numerous and almost completely intact fossils of some of the earliest known whale species that have been found there, including the 20-meter-long *Basilosaurus* and the smaller *Dorudon* (plate 7). That these animals are from an early stage in whale evolution is clear from some of the features that are preserved, such as their teeth, which closely resemble those of the terrestrial animals from which they evolved, and the remnants of hind limbs, anatomical features that have since been lost as the evolution of the whale has continued. A range of other animals has also been preserved in the strata of the Valley of the Whales: sea cows, crocodiles, turtles, and sharks, together with aspects of the habitat in which these animals flourished, including fossilized mangroves similar to those discussed in the last chapter (fig. 4.4).

Bahariya and Farafra Oases—A Geological Wonderland

As the Eocene seas continued to retreat to the north, the landscape that emerged across Egypt and neighboring areas was very different from that of today. Neither the Red Sea nor the Nile Valley existed at this time, and Egypt's northern coast lay at the general latitude of Asyut (fig. 1.1). The climate of this period was also very different, and at this time we see evidence for some of the first identifiable rivers in Egypt. The existence of these ancient rivers was initially revealed by the presence of a large area of coarse rounded gravels in the desert to the west of Faiyum. These gravels have been dated to the Oligocene (36 Mya or 11:50 p.m.) and have been interpreted as the remains of a huge river delta that was much larger than the modern Nile Delta. The fossilized remains of trees within these deltaic deposits indicate that these ancient rivers flowed through a forested landscape and that occasionally the rivers flooded, carrying trees and other remains downstream to be deposited in the delta sediments. In addition to fossilized trees in the ancient delta, fragments of petrified wood can be found today scattered across

Figure 5.1. Well-preserved fossilized wood in the area west of the Faiyum. Although this may look like part of a recently felled but desiccated tree, all the organic material has been replaced by rock-forming minerals, giving specimens like this the cold, heavy feel of stone.

large parts of northern Egypt (fig. 5.1), probably associated with widespread volcanic activity that continued until the early Miocene (fig. 2.1). Groundwater heated by this volcanic activity and carrying large concentrations of dissolved minerals is likely to have caused the petrification of large areas of forests that had survived beyond the lava flows. These huge basaltic eruptions affected large areas of the Western Desert and extended toward the Sinai Peninsula, and as lava poured out across the land surface, often from fissures in the crust, it quickly cooled to form fine-grained igneous rocks, often with a distinctive six-sided columnar structure similar to that seen at the Giant's Causeway in the north of Ireland. Since the formation of these Egyptian basalts, much of the land surface has been heavily eroded, leaving dramatic outcrops of columnar basalt as prominent caps to isolated hills in the area around Bahariya (plate 8).

These volcanic events of the Cenozoic may also be responsible for the formation of one of Egypt's more bizarre geological features, the Crystal Mountain, which was discovered during the construction of new sections of road between the Farafra and Bahariya oases. During these roadworks, a cutting through a low-lying limestone ridge exposed a remarkable fantasia of large-scale crystal growth (plate 9). Although views differ as to the precise origins of this feature, it seems likely that Egypt's Crystal Mountain formed when hot, mineral-rich groundwater entered a former cave system or subvolcanic chamber. As these fluids cooled, crystals began to grow from their mineral content, and as cooling occurred at relatively slow rates, the crystals reached impressively large proportions. There is certainly evidence for a great deal of heat associated with the formation of Crystal Mountain, which in places appears to have altered the rocks that surround the crystalline mass.

Although volcanic activity largely ended in the Miocene, there are indications that its effects may continue to be felt in the Western Desert today. In chapter 4, we discussed the Nubian Sandstone aquifer, a vast groundwater resource contained within the permeable Nubian Sandstone. In the Western Desert, this aquifer lies close to the base of the oasis depressions, and the springs and wells that bring groundwater to the surface yield extremely pure, cool, fresh water. In some cases in the Bahariya Oasis, however, these springs are hot, suggesting that a residual source of geothermal heat exists at depth.

Lost Rivers

As we have already seen, the presence of a large area of rounded gravel in the Western Desert led to the identification of an ancient river delta, the downstream end of a long-lost major river system that predated the Nile by many millions of years. Other than the rounded gravel of the former delta, however, there was little other significant evidence to support the presence of what must have been a major landscape feature. In the 1980s, NASA launched a series of space shuttle missions that were tasked with mapping parts of the Earth's surface using a sophisticated radar mapping unit carried in the shuttle's payload bay. The advantage of radar mapping is that it can "look" beneath the surface covering of sand (in desert regions) or ice (in polar latitudes) and reveal the shape of the underlying bedrock. When NASA analyzed the radar information for North Africa, a network of former river valleys was revealed, hidden beneath the drifting sands of the Great Sand Sea (fig. 1.1). Further shuttle overflights established that these buried valleys were part of an extensive river system now called the

Gilf River, which had been much larger than the modern River Nile. The discovery of the former Gilf River, with valleys that extended from Libya in the west across to Dakhla Oasis in the east, provided further confirmation that the Egyptian landscape of the Cenozoic was very different from that of today. The shuttle radar-mapping missions also helped solve one of the other major problems of Egyptian geology: where had all the sand in the Great Sand Sea come from?

The Great Sand Sea is an enormous area of deep sands and huge dune systems that occupies large parts of the Western Desert, west of the oases and north of the Gilf al-Kebir (fig. 1.1 and plate 10). For many years, researchers had been unable to work out why such enormous quantities of sand were concentrated in this particular area of the desert. Thanks to the shuttle radar-mapping missions, it is now evident that over millions of years, the huge Gilf River system eroded the limestone and sandstone landscape through which it flowed. This vast sediment load was transported downstream and deposited into the shallow waters that lay offshore. At this time, however, North Africa was experiencing significant changes in its regional weather patterns, which culminated about 10 million years ago with the onset of more arid conditions across the region and the establishment of today's dominant northern winds. As regional uplift continued and Egypt's coasts progressively advanced to the north, the vast deposits of sand that had been carried offshore by the Gilf River became exposed and began to dry out in the newly established arid climate. Under the influence of the now-dominant north wind, the countless individual sand grains that had been carried downstream by the Gilf River were driven south, back onto land, eventually to fill and then bury the valleys of the Gilf River. By the time this process had ended, the once majestic Gilf River and its deeply cut system of valleys had been buried in sand, hidden from all but the prying eyes of the shuttle radar. The development of the Gilf River system and its fate is a very dramatic illustration of the often cyclical nature of geology. It is remarkable to think that it required such advanced equipment as the space shuttle to shed light on this fundamental aspect of the evolution of Egypt's landscape.

The Red Sea

As regional uplift continued through the Oligocene, large-scale tensions developed in the African landmass and beyond. This resulted in the development of a series of rifts and faults within the Earth's crust, the most significant of which can still be seen today in the steep-sided canyon of the African Rift Valley. In the Miocene (about 15 million years ago, or 11:55 p.m. on our 24-hour clock—fig. 2.1) another of these deep rift valleys was breached at its southern end by waters from the Indian Ocean, with this single catastrophic event leading to the formation of the Red Sea. The Red Sea was initially much narrower than it is today, however; tectonic forces continue to push the flooded rift apart, currently by as much as 1 centimeter each year.

During the formation of the Red Sea Rift, adjacent land areas had been thrust upward into a series of young mountain chains. On the Egyptian side of the Red Sea, these mountains were to become the Red Sea Hills, which, despite the rather unassuming name, are a high and very dramatic mountain range (fig. 1.5). The jagged nature of the Red Sea Hills is a clear indication that they were formed relatively recently and is consistent with what we understand about their origins in the geologically recent Oligocene Epoch (fig. 2.1). When considering this phase of Egypt's geological evolution, it is important not to confuse the relatively recent tectonic events that led to the uplifting of the Red Sea Hills with the considerably more ancient Precambrian origins of the strata of the mountains themselves. As we saw in chapter 3, Egypt's Precambrian Basement consists of ancient strata that were formed over half a billion years ago. Given their great age and the many earth-forming events they have endured, Egypt's Precambrian strata have developed a rich mineral content that, prior to the opening of the Red Sea, was buried beneath layers of Mesozoic sandstones and Cenozoic limestones. It was not until the Red Sea Hills were uplifted in the Oligocene and subsequent erosion stripped away the overlying limestones and sandstones that the ancient strata of the Precambrian Basement were exposed (plate 11). The mineral-rich Precambrian strata of the Red Sea Hills have therefore been buried for hundreds of millions of years, and in geological terms have only recently been uncovered. I consider that the great mineral wealth of Egypt, and particularly the mineral wealth of the Red Sea Hills, was one of the major influences on the development of the pharaonic state as it emerged along the banks of the Nile some 5,000 years ago.

Plate 11 also illustrates another important geological concept, the unconformity. Unconformities are lines of contact between rock units of greatly different ages and represent gaps in the geological record. Unconformities are very common in geology and represent periods of time either when there was no deposition or when the area of interest was dominated by erosion. Whatever the cause, the rocks that are missing from the geological record and are marked by an unconformity can represent gaps that can be hundreds of millions of years long. The unconformity shown in plate 11 represents an enormous gap in time and is associated with the general absence of Paleozoic strata in Egypt that was discussed in chapter 4. If, conservatively, we assume the lower-lying, dark basement rocks shown in plate 11 are from the very end of the Precambrian (about 560 million years ago, fig. 2.1) and the overlying Nubian Sandstones are from the very earliest part of the Mesozoic (about 240 million years ago), the unconformity shown in plate 11 represents a period of over 300 million years, a vast period of time that is longer than the entire Mesozoic and Cenozoic eras combined (fig. 2.1).

The formation of the Red Sea had implications far across the Egyptian landmass. By the middle of the Miocene (about 15 Mya, or 11:55 p.m.), drainage from the uplifted Red Sea Hills had led to the formation of another major river in Egypt: the Qena River, which, unusually, drained to the south into Africa's heartland. To the west was the Gilf River, and along Egypt's northern coast were a number of lesser rivers, all of which drained to the north into the Tethys Sea (fig. 5.2). This however, was a period in which sea levels in the Tethys were continuing to fall, and by the end of the Miocene (about 5 Mya, or 11:57 p.m.), it appears that most of the Tethys basin had dried out.

All rivers have what is referred to as a base level, which is the lowest point in the river system and is usually the point at which the river meets the sea. By the time that the Tethys had dried out, the

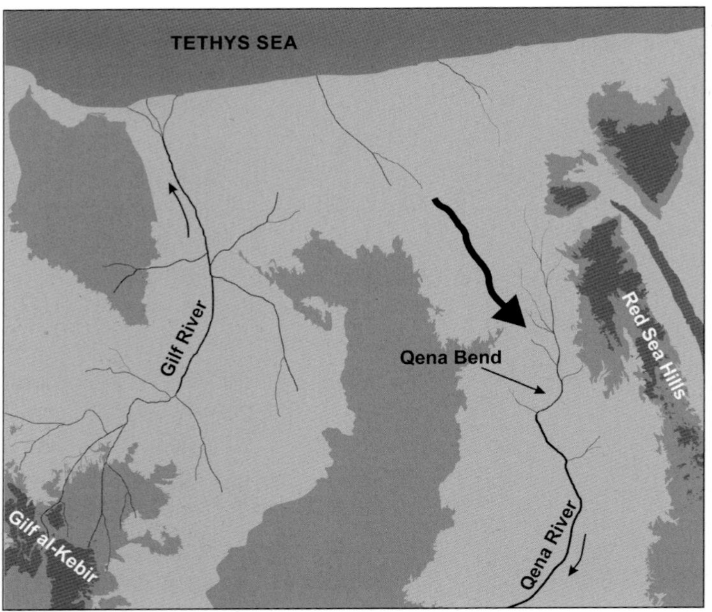

Figure 5.2. Highly conjectural representation of northeast Africa in the late Miocene, showing the Gilf River draining north and the Qena River draining south. (See the text for the significance of the broad arrow.)

base level had become substantially lowered, rejuvenating Egypt's northern-flowing rivers and causing them to both straighten and deepen their channels for considerable distances inland from the coast. As they eroded inland, these rejuvenated rivers exploited weaknesses in the bedrock, most likely associated with patterns of geological faulting caused by the opening of the Red Sea. Although figure 5.2 is highly conjectural, one of the rejuvenated northern rivers must have followed the approximate path shown by the broad arrow, and as it continued to advance southward, it cut an over-deepened canyon toward the Qena River and the area we refer to today as the Qena Bend: that wide sweep of the Nile north of modern Luxor (figs. 1.1 and 5.2). With the headwaters of the northern river approaching the valley of the mature Qena River, the scene was set for one of the most significant episodes in the history of the Egyptian landscape, which, when compared with the long drawn-out backdrop of geological time, must have taken place in the blink of an eye. At a certain point on the Qena Bend, the headwaters of the rejuvenated northern river broke through into the Qena Valley, causing the enormous flow of the Qena River, which had flowed to the south for many millions of years, to reverse and begin discharging to the north. It is likely that at the point of river capture, a huge waterfall developed, with the water of the Qena River cascading over the falls into the deeply scoured northern channel. As this colossal waterfall continued to cut southward into the continental landmass of Africa, a new, deeply cut river canyon was formed, extending as far south as Aswan. At Aswan, the Eonile Canyon, as this enormous erosion feature has been called, did not follow the present course of the river, but bypassed the shallow granite outcrop of the First Cataract (fig. 1.3) to follow a more easterly route before rejoining the river further to the south. Eventually, the Eonile system reached some sort of equilibrium, but not before a canyon had been ripped through the Egyptian landmass that was 2.5 kilometers deep and longer than the Grand Canyon in the United States.

Although we have not yet reached the end of the Cenozoic, it is appropriate to end this chapter at this point in the development of the Nile system. As we will explore in the following chapter, the towering walls of the late Miocene Eonile Canyon have constrained the subsequent development of the river system we know today as the River Nile. However, a great deal has happened between the formation of the Eonile Canyon and the far smaller and relatively sedentary river that we are familiar with today.

CHAPTER 6
THE EVOLUTION OF THE RIVER NILE

The Fate of the Eonile Canyon

We ended the last chapter about 5 million years ago in the late Miocene. The Tethys basin had dried out, rejuvenating the rivers that drained the northern areas of the Egyptian landmass. With renewed energy and vigor, one of these northern rivers advanced southward away from the coast, carving out a deep channel in the landscape of northern Egypt until it eventually broke through into the valley of the ancient Qena River, at a position on the Qena Bend, north of modern Luxor. By the time the consequences of this tumultuous river capture had unfolded, a vast canyon system had developed, running for more than 800 kilometers from Egypt's northern coast as far south as Aswan. Geophysical investigations have shown that beneath Cairo, this massive Eonile Canyon extends to a depth of over 2,500 meters below sea level. The Grand Canyon, to which the Eonile is often compared, is a "mere" 1,800 meters deep. If you have only just read the last chapter, you may be wondering why I am repeating this, but given the scale of what was occurring in Egypt at this time, it *is* worth repeating.

The moment that the crumbling walls of the northern river broke through into the mature valley of the Qena River must have been a truly awesome event. We can only speculate on precisely what happened as the huge flow of the Qena River slowed to a halt before reversing to flow north. It is also hard to imagine the size of the waterfall that developed at the point of river capture. This waterfall probably dwarfed what is today's highest waterfall, the Angel Falls in Venezuela, which is a little under 1,000 meters high. From this time onward, although the Nile was to evolve through five distinct phases, the river north of Aswan was destined to be confined by the steep walls of the Eonile Canyon.

Today, the Eonile Canyon walls are an important and characteristic feature of the Egyptian landscape and can be seen along much of the Nile Valley. In many areas they are set back some way from the Nile or from the sites we tend to visit most frequently, and so they tend to get overlooked. There are locations, however, where the spectacle of the canyon walls cannot be escaped, such as Deir al-Bahari (plate 2), where the towering cliffs provide a dramatic backdrop to the ancient temples and tombs that were built there. What is perhaps seldom understood is that the cliffs that are visible today represent only the upper 10 or 20 percent of the walls of the Eonile Canyon; the lower 2,000 meters or so lie buried beneath the sediments of the River Nile. Perhaps more than any other, this simple fact emphasizes the enormous scale of the Eonile Canyon.

One River, So Many Names: The Eonile, Paleonile, Protonile, Prenile, and Neonile

For many years, geologists were puzzled by the findings of boreholes that had been drilled into the deep sediments that filled the Eonile Canyon. Beneath layers of river sediments, these boreholes revealed that marine conditions had existed in the Nile, and given the thickness of these marine deposits, the canyon had been occupied by the sea for a considerable time. How could marine conditions have existed in a freshwater river valley?

At some point in the Pliocene (fig. 2.1), the waters of what was to become the Atlantic Ocean breached the Straits of Gibraltar and the largely dry Tethys basin was deluged, heralding the modern Mediterranean Sea. Having rapidly filled the former Tethys basin, this inundation surged along the deep Eonile Canyon, transforming it into an arm of the sea that extended as far south as Aswan (fig. 1.1). These conditions lasted for millions of years, during which the marine sediments that had puzzled geologists were deposited. Eventually, conditions in the canyon began to change. Although the reasons for these changes are unclear, they were most likely caused by falling sea levels in the Mediterranean together with the increasing impact of a number of rivers that drained northeast Africa at this time and discharged into the Eonile. Gradually, marine conditions gave way to brackish and then to freshwater environments, bringing an end to the only marine phase of the River Nile. By the time that freshwater conditions had become re-established in the Eonile, the lower half of the canyon had become filled with marine sediments.

As discussed in greater detail by Rushdi Said (see Further Reading), the Paleonile was the first of a series of post-Eonile freshwater rivers that flowed within the Eonile Canyon and existed from the late Pliocene and into the Pleistocene (from about 3.5 Mya to 1.8 Mya—fig. 2.1). The red-brown clays, silts, and sands that are typical of the Paleonile suggest that it was fed by tributaries that drained the area of today's Eastern Desert, with little, if any, contribution from outside Egypt. The consistent nature of the Paleonile sediments indicates that Egypt was not as arid as it is today, with a much wetter climate and more widespread vegetation. The Paleonile appears to have been a huge river and by the end of this phase, the Eonile Canyon had become completely filled with freshwater sediments, overlying the marine sediments that had been deposited in the earlier Eonile phase.

The end of the Paleonile phase, like the end of most of the subsequent phases of Nile evolution, was marked by a period of aridity with greatly reduced rainfall. Despite this, the interval between the Paleonile and the subsequent Protonile phase was a period of great erosion across Egypt, with evidence suggesting that the generally arid conditions were interrupted by brief periods of torrential rain. It was during this period that the majority of Egyptian alabaster (correctly known as travertine) was formed in Egypt. Travertine is a highly decorative, pale or even translucent stone, often with a concentric banded appearance, that was adopted for a variety of important uses throughout the pharaonic period. The travertine deposits of the Paleonile/Protonile interval were formed in pools of standing water that were left behind following the sporadic heavy rains. As these pools dried out in the prevailing arid conditions, almost pure calcium carbonate was deposited in thin layers around the algae and plants that thrived in the pools. As these layers built up over time, they developed into travertine deposits up to 7 meters thick.

The arid events that brought an end to the Paleonile phase of the river were superseded about 650,000 years ago by the development of the relatively short-lived Protonile. The extensive rounded, coarse quartz and quartzite gravels associated with this phase of the Nile can still be seen across many parts of the Nile Valley and indicate that much of the runoff that drained into the Protonile was from the Western Desert. The extensive distribution of these coarse gravels suggests that the Protonile was a very competent river fed by significant rains that were sustained throughout most of the year. Although it was still largely confined within the walls of the Eonile Canyon, the Protonile appears to have followed a more westerly channel than the earlier Paleonile, and it was probably during this phase that a sustained period of repeated Nile floods led to water escaping the limits of the Eonile Canyon. By flowing along former drainage channels in the south of Egypt, extensive low-lying areas of what is now the Western Desert were flooded to form watercourses and lakes. On a geological timescale, the surface water features that were produced by this flooding were relatively short-lived. However, Ted Maxwell and his colleagues have identified evidence for a particularly large lake, referred to as Lake Tushka, that may have occupied an area south of Kharga Oasis (see Further Reading and plate 12).

Like the Protonile, the next phase of Nile development, the Prenile, was also a large river, significantly larger than the Nile today. Although the dating is uncertain, the Prenile phase is thought to have lasted from about 700,000 to 200,000 years ago, with the deposits laid down during this period being characterized by uniform sands and gravels. Most of these deposits appear to have been eroded from landscapes that lay beyond Egypt's boundaries, suggesting that Egypt had a relatively dry climate at this time. For the first time, the deposits laid down by the river include material eroded from the Ethiopian Highlands. This phase of the Nile played a major role in shaping the modern landscape of Egypt, with the majority of the deposits that form the Nile Delta laid down by the Prenile. It was also probably toward the end of the Prenile that waters from the Nile Valley first broke into the Faiyum basin (chapter 9).

Unlike the other phases of the Nile, the Prenile ended not with a time of aridity but with a dramatic change in conditions that saw rains over Egypt, which were sustained throughout most of the year. Sediments associated with this phase of the river contain much evidence for the erosion of basement strata from the Eastern Desert (chapter 7), indicating that at this time, the Nile Valley was dominated by runoff from within Egypt, with little contribution from adjacent areas. It was only in the final stage of Nile evolution (the Neonile—from approximately 120,000 years ago) that the connections with drainage from Ethiopia were re-established, though on a much smaller scale than had previously been the case.

Technically speaking, it is the Neonile that flows through Egypt today, and as Herodotus famously stated, it was a major catalyst for the development of the cultures of ancient Egypt. As we have seen, the river that acted as the main life-giving artery of the pharaonic civilization was the last in a series of phases through which the evolving Nile system passed and is the result of an often dramatic series of earlier geological events. These include river capture at the Qena Bend, connection with drainage systems initially to the east and west of the river, and then, ultimately, connections with drainage systems from Ethiopia to the south. Although each of these events played a distinct role in shaping the Neonile, it was the connections with Ethiopia that were perhaps the most fundamental. As we will discuss shortly, the drainage links with Ethiopia provided the conditions for the life-giving annual inundation that allowed for the development of sophisticated cultures along the banks of the river. It is important to recognize, however, that even slight differences in the way any of these key events unfolded could have led to a very different pattern of evolution for the River Nile, and ultimately, the pharaonic culture that is so widely celebrated around the world may never have developed.

Climate Change

The various stages of the River Nile seem to have developed in response to major climatic changes. The transition to the current, generally arid conditions of northeast Africa is thought to have begun about 10 million years ago, but the Earth's climate is never static, and despite the predominantly arid conditions of the last several million years, remarkable alterations in the climate have occurred. Today a belt of low pressure (the Inter-Tropical Convergence Zone, ITCZ) circles the equator. However, as recently as 12,000 years ago, the ITCZ was located further north, resulting in a relatively brief period of less arid conditions over Egypt known to archaeologists as the Holocene Wet Phase. The more benign climate of the Holocene Wet Phase allowed vegetation to expand to the point that areas of Egypt's deserts may have resembled the savannah of modern Kenya (fig. 6.1), with seasonal lakes and drainage systems. In addition, the fertile areas of the Nile Valley and the regions of Faiyum and the oases will have been more extensive than they are at present. These changes to the natural environment are thought

Figure 6.1. Today's Kenyan savannah. During the height of the Holocene wet phase between 12,000 and 6,000 years ago, areas of Egypt outside the Nile Valley may have looked like this.

to have lasted until about 6,000 years ago and, as we will discuss in later chapters, encouraged a relatively brief period of human repopulation of large areas outside the Nile Valley.

The Current River Nile

Although much smaller than any of its earlier phases, the Neonile is one of the modern world's longest rivers, flowing through Tanzania, Uganda, Ethiopia, South Sudan, Sudan, and Egypt on its journey north to the Mediterranean (fig. 1.1). Since the end of the Holocene Wet Phase, the River Nile has been unusual in that for over 2,000 kilometers, from its confluence with the Atbara River in northern Sudan to the base of the Nile Delta at Cairo, the river has been confined within a single channel without any tributaries. Although for most of its length this single channel is broad and readily navigable, in places it is constrained by the geology of the areas through which it flows. We encountered the most northerly of these geological bottlenecks when discussing the First Cataract at Aswan in chapter 1, where an outcrop of hard granite disrupts the otherwise deep and wide channel of the Nile (fig. 1.3). A series of five similar cataracts have been identified further south along stretches of the river in Egypt and Sudan, although two of these are now submerged beneath the waters of modern reservoirs.

Three hundred kilometers south of the confluence with the Atbara and south of the Sixth Cataract, near the Sudanese capital of Khartoum, two major tributaries combine: the White Nile and the Blue Nile. The White Nile is the longer of these rivers; it rises in the equatorial highlands of the Great Lakes area of Central Africa and has a relatively constant flow throughout the year. It is the contribution from the Blue Nile, however, that gave the Nile in Egypt one of its defining characteristics, a feature that we have already considered and is not only important for our story, but arguably is of huge importance for the history of human culture. This is a *big* statement, but one I'm confident is quite justified.

Flow along the Blue Nile is highly seasonal and is swollen enormously when monsoon rains fall over the upstream catchment in the Ethiopian Highlands. These summer rains wash soils from the mountains into the headwaters of the river, with this turbid flow channeled through the confluence with the White Nile at Khartoum and north along the Nile. Before the construction of modern dams, the swollen river would enter Egypt at the First Cataract at Aswan and would surge along the narrow Nile Valley, through the branches of the Nile Delta, eventually discharging into the Mediterranean. The volume of this floodwater was too great for the normal channel of the river in Egypt, and as a result the late-summer flow would overtop the river's banks, causing widespread flooding across the broad Nile floodplain. Under different circumstances, this flooding could have been catastrophic, washing away settlements or agricultural areas, but the floods carried with them huge quantities of highly fertile silts that were laid down across the floodplain, producing ideal farming conditions that were embraced by the early cultures of the Nile Valley. Canals were dug to allow the floodwaters to reach as far as possible from the main river channel, and embankments were constructed to retain the water on the land and so maximize the benefit of the sediment that was deposited. From these early engineering works, the cultures of the Nile Valley learned to modify the landscape in which they lived, and it is my view that the skills in engineering that were developed at this time contributed vastly to what was soon to follow: the construction of the pyramids, one of man's greatest engineering achievements (plate 13).

We will examine the development of building in stone in Egypt in chapter 11, but for now will focus on the agricultural wealth that developed from the fertile soils that were laid down by the annual Nile flood. This agricultural wealth was so great that it allowed the early cultures of the Nile Valley to move away from long-practiced methods of subsistence farming. In a subsistence economy, each family aims to grow sufficient produce for its own needs. In some environments, even this can be difficult, but thanks to the benevolent Nile, except in years of abnormally low or high floods, farmers in Egypt could readily produce more than their immediate family needed. The potential to produce an agricultural surplus presented great opportunities, which, as the archaeological record shows, were fully embraced by the people of the Nile Valley.

Most emerging subsistence economies develop skills in small-scale domestic activities such as pottery or weaving. These skills are usually born out of necessity, and the products that are made are generally for the use of the individual families that produce them. In Egypt, however, the capacity to produce an agricultural surplus meant that not everyone needed to farm the land, and individuals, or indeed whole families, could dedicate themselves to other activities. Initially the exchange of excess produce for items such as clothing, pottery, and processed foodstuffs such as bread and beer is likely to have taken place on a relatively small scale, with many families continuing to make products such as these in small volumes for their own consumption. However, the exchange of agricultural products for manufactured goods in Egypt is likely to have stimulated an early trading economy, and as this developed, the consequences for early cultures in the Nile Valley were profound. Rather than simple subsistence economies, the pre-pharaonic cultures of Egypt emerged as communities with a range of skilled craft specialists. As the economy grew, the demand for goods manufactured by these artisans led to markets for nonessential or even luxury goods including wooden furniture, decorative (rather than purely functional) stone vessels, and metal objects, including the ornate inlaid gold jewelry for which the pharaonic civilization is justly renowned (fig. 1.7).

It now seems increasingly likely that the predynastic cultures of Egypt developed largely as a result of widespread trading activity, with trade and commerce, rather than conflict, leading to the emergence of the world's first nation-state. It also seems that once the concept of nationhood had become established across the Two Lands, bureaucracy was quick to follow. Pharaonic bureaucracy required scribes and a system of writing to record the various transactions on which all levels of the state functioned, and out of this need, the beautiful system of hieroglyphic writing developed. The agricultural surplus that could be produced in the Nile Valley allowed the ruling classes to levy taxes, which, in a society without currency, were handed over to the state based on a percentage of the crops or livestock that were produced each year. This system of taxation was so important, and the quantities of agricultural produce

Figure 6.2. Enormous mudbrick vaults built behind the mortuary temple of the New Kingdom pharaoh Ramesses III at Medinet Habu, on the west bank of the Nile at Thebes. These structures, and many others like them along the length of the Nile, were built to store the vast quantities of grain that the pharaonic state took in taxes from ancient Egypt's farmers. The location of these granaries, adjacent to a royal mortuary temple, indicates the importance of these reserves to the pharaonic state.

received as taxes were so significant, that huge granaries needed to be built, often adjacent to other important pharaonic buildings (fig. 6.2). As much as reserves of gold or other precious materials, these reserves of agricultural produce represented the great wealth of ancient Egypt and were either retained for times of famine, were used to "pay" nonagricultural citizens such as soldiers, masons, and builders, or formed the basis for trade with neighboring states.

Despite the Nile being generally benevolent, it is clear that periods of abnormal floods could be disastrous for ancient Egypt. If the floodwaters were too high, the canals and embankments would be overwhelmed and the benefit of the flood would be lost. If there was a succession of low floods, drought and famine would result. As shown in figure 1.6, the pharaonic era has been divided into three great historic periods (the Old Kingdom, Middle Kingdom, and New Kingdom), separated by three lesser, or intermediate, periods. Although the circumstances that brought about the end of each of Egypt's great kingdoms are not fully understood, there are strong indications that the demise of the Old Kingdom may have been the result of famine. To understand the causes of this, we have to go back to the end of the Holocene Wet Phase about 6,000 years ago and the subsequent transitional phase of increasingly arid conditions. As discussed by Jean-Daniel Stanley and his colleagues (see Further Reading), the end of this transitional phase occurred sometime around the end of the Old Kingdom and appears to have coincided with a series of successive low Nile floods. Reserves held by the state may have been sufficient to survive one or two years of the resulting poor harvest, but as low Nile flood levels were sustained, their impact on pharaonic society would have been increasingly severe. With pressures such as this on the society, centralized government may have broken down, and state-run projects, such as the construction of the royal pyramids, would have ceased. It is evident that, thankfully, these environmental pressures eventually lifted, allowing centralized authority to be re-established in the form of the period we refer to as the Middle Kingdom.

It is likely that concerns over abnormally low or high Nile floods may have led to the construction of Nilometers along the river from Aswan to Cairo. We do not know when the first Nilometer was built, but these gauging stations provided points at which the height of each year's Nile flood could be measured (fig. 6.3). These annual flood levels were regarded as so important that during the Old Kingdom they were carved into stone tablets, known today as Royal Annals. The Royal Annals provided a record of the key events during the reigns of a number of pharaohs from ancient Egypt's earliest dynasties, and although no complete Royal Annals have survived, fragments from at least two copies are kept in museums in Palermo (the Palermo Stone), in Cairo, and in London. Most of the events recorded on the Royal Annals fragments are considered to be highly embellished accounts of the principal achievements of each pharaoh. However, alongside the royal propaganda, the Annals record the height of each year's Nile flood. The fact that such important state archives made a point to record annual flood levels is a clear indication of the importance of the Nile inundation to the culture of ancient Egypt.

Today, all of Egypt seems to be focused on the river, from the bustling metropolis of Cairo to the smaller, provincial towns.

Figure 6.3. Looking down inside the highly ornate Nilometer at Roda Island, Cairo, built in AD 861 on the site of an older Nilometer. Nile waters could enter the structure at the base and the level of the annual Nile floods would be recorded on the central octagonal column.

Between these towns, scenes of farming and fishing continue to play out as if they are defying the passage of time, and from the river we see the variety in the landscape through which it flows: open pastures (fig. 1.2) give way to narrow canyons with towering limestone cliffs. Fishing boats, rowing boats, and garishly painted car ferries crisscross the river, underlining the fact that the River Nile *is* Egypt and Egypt *is* the Nile. The close relationship between the Egyptian state and its only major waterway results in a nation that is extremely elongated, a ribbon of green pasture strung out for more than 1,000 kilometers along the banks of the mighty river (fig. 1.1). It is remarkable that the elongated pharaonic state, which relied so much on agriculture and trade, could function so efficiently. Clearly, the easiest means to travel through Egypt was by boat, navigating the Nile's current from south to north to travel from ancient Aswan to Memphis (just south of modern Cairo). To speed up a northward journey, vessels could use oars; however, to return south, it is unlikely that rowing against the current would have been a practical option. For a southbound journey, the ancient Egyptians could rely on yet another accident of nature: the dominant northerly winds that, since the demise of the Gilf River some 10 million years ago (see chapter 5), have blown in from the sea. These winds allowed the use of sails to travel against the river's current; without them, it would have been far more difficult for ancient Egypt to operate as a single, unified realm.

Rivers are never static, and we cannot assume that what we see today reflects conditions in a river valley several thousand years ago. Indeed, understanding the dynamics of the River Nile and the effects this may have for our interpretation of the archaeology of ancient Egypt may be one of the greatest challenges facing modern Egyptology. As an example, we can look at one of Egypt's most visited monuments, the New Kingdom temple of Karnak. The temple structures at Karnak were developed progressively over a period exceeding 1,500 years, from the Middle Kingdom into the Late Period (fig. 1.6), and archaeologists have long known about a series of substantial walls located a short distance in front of the temple complex (fig. 6.4). These walls were built in the Third Intermediate Period, and because today the Nile is about 0.5 kilometers to the west, it had been assumed that they were part of a system of ancient canals and harbors that had been built to connect the sacred site to the river. Recent studies by John Hillier and his colleagues (see Further Reading) have focused on developing an understanding of the ground conditions at Karnak and have established that since development of the Karnak site began, the channel of the River Nile has moved westward. All rivers tend to meander naturally within their valleys, accentuating natural bends and causing the channel to alter its position over time. As the position of a river changes, new areas of land are left in its wake. Not only is it now clear that some sections of Karnak temple have been built over areas of former river channel, it has also been established that the walls in front of the temple are not canal walls as previously believed, but are former river walls. The dynamic nature of the River Nile will not have been restricted to just the Karnak area, and over archaeological timescales, other stretches of the river will have changed in similar ways. The realization that, during and since the pharaonic era, the Nile has been actively meandering between the limestone cliffs of the Eonile Canyon opens up a new and potentially exciting aspect to the archaeology of the Nile Valley. What hitherto unsuspected

Figure 6.4. A section of large masonry wall (arrowed) exposed in front of the main pylons at Karnak temple. Initially thought to have been harbor walls, these are now understood to be river walls and indicate the extent to which the river has moved in the last 2,000 to 3,000 years. Today the River Nile lies over half a kilometer from this location.

archaeological remains have survived, buried in the silt of former sections of the river?

Despite its meandering across its wide sediment-filled valley for many thousands of years, attempts were made to tame the great river in the early twentieth century, with the building of dams at Aswan. The first dam (now known as the Aswan Low Dam) was built in 1902, but it was not until the construction of the High Dam in 1970 that the annual Nile inundation in Egypt was brought fully under control. In addition to impounding water for the generation of electricity, the High Dam allows the flow of the river to be controlled so that a year-round steady supply of water is available for irrigation, without the need for large parts of the floodplain to be underwater for three or four months each year.

I do not feel it is any exaggeration to conclude this chapter by stating that the emergence of the great ancient Egyptian civilization is unlikely to have developed in the way that is so widely celebrated today if it had not been for a series of accidents of nature. Without key geological events such as the capture of the Qena River, the annual rains that washed soils from the Ethiopian Highlands downstream to flood the Nile Valley, or the dominant northerly winds that made two-way river traffic possible, the glories of ancient Egypt may never have emerged. The evidence that has been gathered by countless archaeological teams during decades of excavation in Egypt makes it clear that the emergence of the pharaonic culture owed a great deal to the geology and landscape of Egypt.

It is equally clear, however, that any natural disruption to the generally benevolent conditions could spell near-disaster for the cultures of ancient Egypt. The Old Kingdom was the period in which the much-celebrated pharaonic culture finally emerged and some of the world's most incredible monuments were built. We have seen, however, that difficult climatic conditions, leading to sustained drought in the Nile Valley, may have brought about the demise of the Old Kingdom. Fortunately this was not the end of the pharaonic era, and from the dark age of the First Intermediate Period a new period of great dominion, the Middle Kingdom, emerged, leading subsequently to the wonders of the New Kingdom (fig. 1.6). As a geologist, it is sobering to think that the pharaonic civilization of ancient Egypt was influenced so greatly by the geological processes that, over millions of years, shaped the Egyptian landscape.

CHAPTER 7
THE EASTERN DESERT

As we have explored the landscape of Egypt, we have become familiar with the pharaonic concept of *kemet*, the rich and fertile landscape of the Nile Valley, which is framed by the tall cliffs of the Eonile Canyon (plate 2). Journeying along the Nile, it is easy to overlook the occasional gaps in these cliffs, many of which offer a window into the other aspect of the ancient Egyptian landscape: the arid wastes of *deshret*, the Red Land. Many of the openings in the eastern walls of the Eonile Canyon are the downstream end of a network of wadis, the culmination of a natural drainage system that rose in the Red Sea Hills at a time when Egypt's climate was much wetter than today. Many of these wadis appear to have been present during the Eonile phase of the river (chapter 6) and continued as active tributaries of the Nile at least during the subsequent Paleonile phase, allowing distinctive red-brown deposits to be eroded out of the Eastern Desert and distributed along the Nile Valley. With its watershed among the peaks of the Red Sea Hills, this drainage system is likely to have been influenced by the geological structure of the landscape, with faults and fissures in the bedrock providing zones of weakness for the flowing water to exploit, as it eroded the landscape of what is

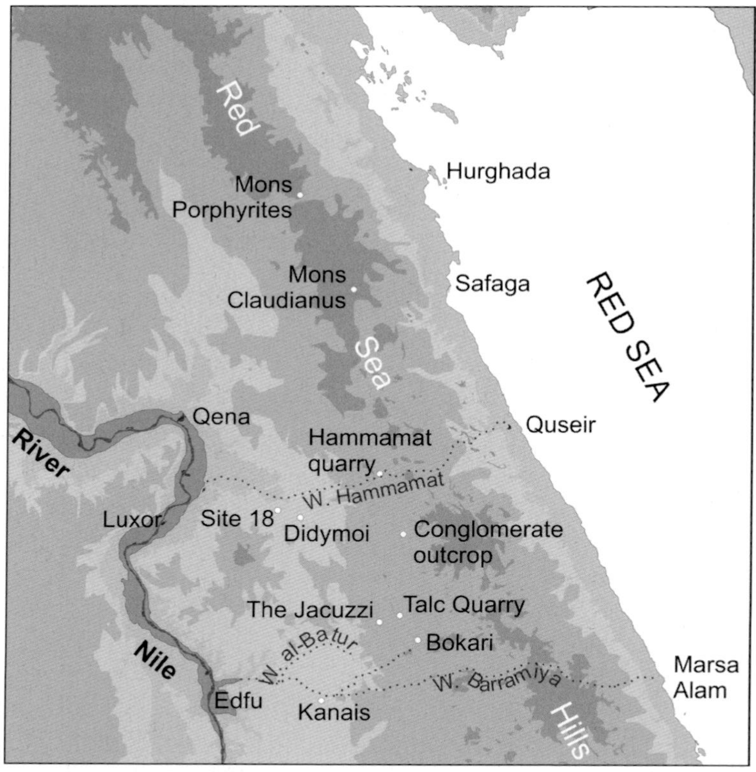

Figure 7.1. Sketch map of the central Eastern Desert showing places referred to in the text.

now the Eastern Desert. Although today most of Egypt experiences a generally arid climate, as a result of the elevated terrain of the Red Sea Hills the Eastern Desert will typically experience more rainfall than other areas of Egypt's deserts. The consequences of this rainfall will be considered shortly, but we should not lose sight of the fact that rain in the Eastern Desert is an infrequent event that lasts for the briefest time, before being dried out by the intense heat of the desert sun.

The character of the landscape in the Eastern Desert is influenced greatly by the underlying geology. In the east, the resilient backbone of Precambrian strata that were thrust upward to form the Red Sea Hills provides a rugged, elevated landscape that descends abruptly to a narrow coastal strip. Between the Red Sea Hills and the Nile Valley, the Eastern Desert is characterized by an extensive plateau into which the wadi systems have been eroded. As the simplified geological map shows (plate 3), north of Edfu, the surface of this plateau consists mainly of Cenozoic limestones (in blue); however, closest to the Red Sea Hills is a zone of outcropping Mesozoic sandstones (in green), which tapers to the north. South of Edfu, the plateau consists entirely of Mesozoic sandstones. Wadis in the limestone areas tend to be relatively broad compared with those eroded into sandstone strata; however, all wadis share a number of characteristic features including wide, flat, sandy floors overlooked by rocky walls that, depending on their alignment, lie in shade for parts of the day (plate 14). Like the Eonile Canyon, though on a much smaller scale, the wadis of the Eastern Desert have become partially infilled. The sediments that fill the wadis tend to be deepest in the downstream sections, closest to the Nile, and were mostly deposited when the climate was less arid and rainfall over the Red Sea Hills led to runoff through the wadi system. Once the area had become fully arid, these fluvial sediments were overlain by drifting sands.

Although the evidence they have left behind is sparse, throughout the historic period and beyond, the Eastern Desert has been thinly populated by nomadic people who were able to survive in this challenging environment only through their intimate familiarity with the desert landscape. As we saw in chapter 1, the inhabitants of the Nile Valley were raised to view the Red Land as a place of chaos, the realm of the dead that was inhabited by powerful and often irascible gods. Despite what would appear to be a cultural aversion to the Red Land, however, there is clear evidence that throughout the pharaonic period, people from the Nile Valley were active in the Eastern Desert. This evidence suggests these desert travelers were mainly drawn by commercial activities, either trade with the Red Sea coast or the lure of the great mineral wealth that lay in the Red Sea Hills. Whatever the motivation for this commercial activity, it is likely that it benefited greatly from the presence of the wadi system. Although the meandering wadis of the Eastern

Desert do not provide the most direct route for travelers, and despite some wadis becoming blocked by large sand drifts, the even, sandy surface of the wadi floors, the gentle gradients, and opportunities to shelter from the direct heat of the sun combined to make desert travel along the wadis a less challenging proposition than attempting to journey over the rocky surface of the elevated plateau above. Viewed from the surface of this plateau, the Eastern Desert looks surprisingly alien: a vast, featureless expanse, flattened by the wind and the fiercely abrasive sand that the wind carries. From the plateau, you tend to look across the incised wadis, often without realizing they are there. The contrast between the exposed open plateau and the relatively benign conditions within the wadis, two key aspects of the same landscape, is startling.

Given the advantages associated with traveling through the wadi system, it seems that from an early time the major wadis became established routes across the desert, and many of these routes are still in use today. Two examples of these ancient wadi routes are given in figure 7.1, though many more were in use throughout the area's rich past. The most southerly route shown starts at Edfu in the Nile Valley and follows Wadi Abad and Wadi Barramiya, before reaching Marsa Alam on the shores of the Red Sea. Today, this route and the second, more northerly, identified route through Wadi Hammamat are paved with asphalt and dotted with occasional truck stops and mobile-phone masts, but these do little to tame the wildly dramatic landscape of the Eastern Desert. Wadi Hammamat is renowned for the *bekhen*-stone quarries that lie about halfway along the wadi road, in an area where erosion has cut deeply into the towering foothills and the dark color of the bedrock adds a brooding air to the landscape (fig. 7.2). The summits of the highest peaks are strung with the ruins of ancient towers built from the abundant local stone and used for security as well as a system of line-of-sight communication from one end of the wadi to the other. We will return to Wadi Hammamat in chapter 10, as this area of the Eastern Desert was not only the main ancient source for pharaonic *bekhen*-stone, but was also exploited extensively throughout the pharaonic and Roman periods for gold.

Figure 7.2. Wadi Hammamat in the heart of the Red Sea Hills.

Water in the Desert

Although influenced primarily by geology, the landscape of the Eastern Desert has also been shaped by the ever-changing climate, with only occasional evidence for the influence of man. At places such as Edfu (fig. 7.1), modern irrigation schemes transform the broad downstream sections of wadis into commercially important areas of cultivation, but as you travel east away from the Nile, nature rapidly gains the upper hand. For such a generally arid environment, it is surprising how many trees and other areas of vegetation are present in the Eastern Desert. Trees can be surprisingly large and appear remarkably healthy (plate 15) and are often accompanied by stands of low thorny bushes. These patches of vegetation tend to follow a meandering path along the wadi floor, often toward the center of the wadi channel, a characteristic distribution which provides a clue to understanding their key to survival.

It has been established that since Roman times, significant rains have fallen over the Eastern Desert, on average about once every 50 or 100 years (see Valerie Maxfield and David Peacock in

Further Reading). Although smaller rain events are likely to have occurred more frequently than this, they will have left little trace. During one of my visits to the granite quarry of Mons Claudianus (chapter 10), I was surprised to find clear evidence for recent rainfall. Although the surface of the wadi was dry, there were areas in which the sands had been disturbed by flowing water. Only a few centimeters below the surface, the underlying sediments were wet. The wadi sediments in the Eastern Desert were deposited by watercourses that were active during the earlier phases of the River Nile (chapter 6). Within these generally sandy deposits will be coarser layers of gravel that were laid down during occasional, geologically ancient floods. Modern rainfall will still be influenced by these ancient drainage systems, with infiltrating rainwater collecting in layers of gravel that may be present in the wadi infill. Although modern rainfall is infrequent and in most cases these gravels will retain water only seasonally, these minor gravel aquifers provide a valuable resource for both the hardy desert vegetation and experienced desert explorers.

At the meeting point of several wadis about 60 kilometers east of Luxor is one of the few surviving examples of ancient habitation to be found in the Eastern Desert: the Roman fort of Didymoi (more correctly referred to as a praesidium). Since falling into disuse, the stone walls of the fort have crumbled, but enough of a pair of small drum towers flanking the ancient gateway survive to illustrate the original defensive nature of this structure (fig. 7.3). Didymoi was no doubt established at this location to protect the well that lies within the enclosure, and just like the roots of the hardy desert trees, the well at Didymoi is likely to tap into a gravel aquifer that lies beneath the wadi floor. Even at the confluence of several large wadis such as at Didymoi, however, the year-round availability of water could not be guaranteed. To address this, cisterns had been built between the well and the northern wall of the praesidium. Although today these cisterns are largely filled with sand, exposed sections of the upper walls reveal the white plaster that was used to line the cistern to prevent the precious water from seeping away. The remaining space within the praesidium was divided into a series of buildings, in some of which ancient plaster still clings to the walls, and the smoke-blackened remains of brick fireplaces, so vital for fighting the cold of the desert nights, are easily identified. It is likely that many of the buildings within the fort were used for administration, but there were also designated living spaces and a bathhouse. Perhaps the most remarkable feature to survive at Didymoi is the site's midden, an area just outside the walls where waste from the praesidium was thrown. Over the centuries, the midden is likely to have been picked over by birds or desert foxes so that today, all that is left is a low mound of pottery, mostly fragments but with one or two intact vessels. Over time, the midden will continue to slowly erode to dust, to be carried off by the desert winds.

When considering natural processes such as the development of river systems across a landscape, we tend to think of transitions occurring gradually over prolonged timescales, but this may not always be the case. Dramatic evidence for abrupt changes to the hydrology of the Eastern Desert can be found at a site called the Jacuzzi (fig. 7.4). As the name suggests, the Jacuzzi is a deep, round, bath-like hollow, set back a short distance from one of the wadis that cuts through the sandstone landscape of the Eastern Desert. The smooth sandstone surfaces adjacent to the Jacuzzi indicate that a narrow and fast-flowing watercourse once ran through this area (arrowed in fig. 7.4). At some point, however, local earth

Figure 7.3. The main gate of the Roman praesidium at Didymoi.

movements affected this part of the Eastern Desert, raising the watercourse above the nearby wadi floor. In this way, a classic waterfall developed as the watercourse flowed over the newly formed falls and the cascading water cut a deep basin in the sandstone beneath. In a global context, the processes of waterfall- and plunge-pool formation that we see at the Jacuzzi are relatively common, and given sufficient time, waterfalls such as this will cut a deepened channel, eroding upstream until eventually all traces of the falls are lost. At the Jacuzzi, however, before the watercourse could cut back its newly deepened channel, the climate of the area changed and the watercourse dried up. What we are left with at the Jacuzzi is the geomorphology of a typical small waterfall in a very untypical state, without any water.

The Red Sea Hills

The Jacuzzi lies close to the foothills of the Red Sea Hills, and as you travel further east through the wadi system toward the ancient Precambrian mountains, it becomes increasingly clear

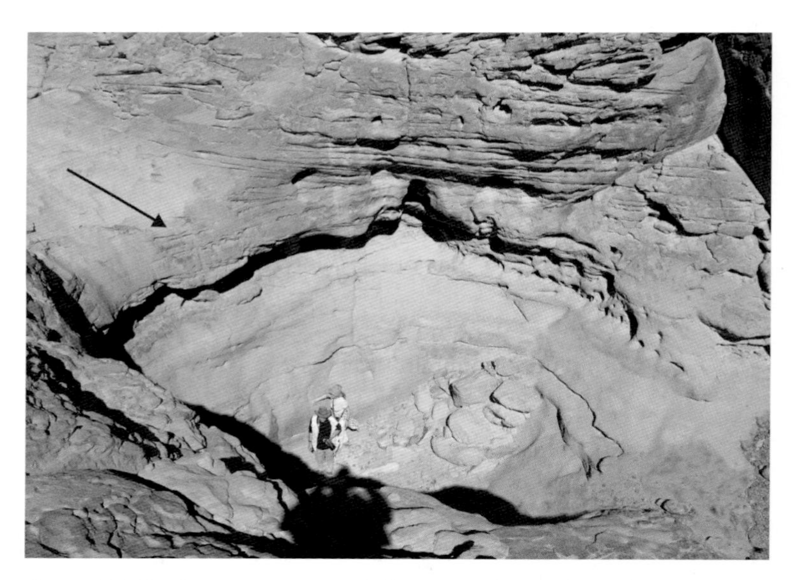

Figure 7.4. Exploring the Jacuzzi. The arrow (upper left) shows the direction of flow of the watercourse that eroded this deep basin.

that the landscape is changing. The best way to appreciate this changing terrain is to find an elevated vantage point, which in the heat and desert scree can be quite a challenge. Views such as that shown in plate 16 reward the effort, however, as they emphasize the scale of the Eastern Desert, with individual cone-shaped weathered peaks stretching away in the distance to merge with the towering coastal mountains. As discussed in chapter 5, the Red Sea Hills are a relatively young mountain range, formed during the opening of the Red Sea Rift in the geologically recent Oligocene Epoch (fig. 2.1). Initially, this uplifted landscape would have looked quite different, dominated, like most of the rest of Egypt, by exposures of Mesozoic sandstones and Cenozoic limestones. Subsequent widespread erosion of the area, however, led to the removal of these younger rocks to expose the underlying Precambrian strata, with only isolated sandstone and limestone outliers remaining along the narrow strip between the Red Sea Hills and the coast (the limited areas of blue and green shown in plate 3). Because the erosion that removed these sandstones and limestones was a relatively recent geological event, the youthful Red Sea Hills retain a rugged and dramatic appearance (fig. 1.5). This relative youthfulness also explains why the exposed Precambrian strata retain much of their mineral wealth. Having been exposed only recently, the ancient mineral-rich rocks of the Red Sea Hills have not been subjected to prolonged weathering and erosion, which may have removed much of the mineral content. It was undoubtedly this mineral wealth that, from the Predynastic Period through to the Greco-Roman period and beyond, drew expeditions out of the Nile Valley and into the harsh desert wastes.

Even with the benefit of modern four-wheel drive, it is not possible to cross the high mountain peaks of the Red Sea Hills, so to reach the Roman quarries of Mons Claudianus or Mons Porphyrites (fig. 7.1), it is necessary to follow the age-old wadi roads through the mountains, perhaps passing through Wadi Hammamat (fig. 7.2), before reaching the shores of the Red Sea. With the inviting silver-blue waters to your right, the wide modern coastal highway whisks you north as far as Safaga, before you turn inland to head once more through the narrow wadi roads toward the distant Nile

Valley. The quarries at Mons Porphyrites were worked during the Greco-Roman period for the purple-colored igneous rocks that outcrop there. Purple was the color that symbolized Roman imperial power, and this unique stone was highly prized throughout the Roman world. We will explore the Roman quarries of Mons Claudianus in chapter 10. Aside from these major quarry sites, there is other evidence for habitation in these remote areas of the Eastern Desert. About 1 kilometer southwest of the main settlement at Mons Claudianus are the remains of buildings known collectively as the Hydreuma. The Hydreuma may have formed the center of operations for some of the earliest phases of Roman granite quarrying in the area, and like most other buildings in the Eastern Desert, was built using blocks of local stone without mortar. The buildings appear to be clustered along either side of a natural erosion gully, and among the surviving features are a series of plaster-lined cisterns, similar to those at Didymoi. Unlike at Didymoi, however, these cisterns were not filled from a well; no well has yet been found at this site. The location of the cisterns suggests they were used to directly capture and store water that periodically flowed out of the gully following sporadic rainfall, an arrangement that not only illustrates the variety of techniques that were required to survive in the desert, but also the ingenuity of the people that ventured into this arid landscape, over 2,000 years ago. About 1.5 kilometers southwest of the Hydreuma is a settlement that illustrates a very different way of desert life. Dating probably to the fifth or sixth centuries AD, the site known simply as the "Enclosure and Tower" has been identified as a possible early monastery. The enclosure surrounds a well, and although often referred to as a possible water tower, the freestanding structure is more likely to have provided a refuge from attack. The single entrance to the tower is some distance above ground level and a ladder would have been needed to gain access. Once inside, the tower occupants could simply have withdrawn the ladder, making access difficult for attackers.

Eastern Desert Rock Art

The Eastern Desert gave me my first experience of a phenomenon that is widespread throughout Egypt, Africa, and beyond: rock art.

The rock art of Egypt's desert areas only came to the attention of Egyptologists in the first half of the twentieth century, when explorers started to report images of people, animals, and perhaps most strangely in the desert, boats that had been scratched or pecked into the rocky walls of the wadis. Although the concept of rock art in the Eastern Desert was largely new to the West, some elements were familiar. For example, many of the boats that were depicted had deeply curved hulls, rows of oars, and decorated prows (fig. 7.5) that resembled those painted on pots unearthed at predynastic and Early Dynastic sites in the Nile Valley. With what may be ostrich feathers in their hair and with upraised arms, many of the representations of human figures found in the Eastern Desert (fig. 7.6) were also similar to those known from predynastic and Early Dynastic ceramics.

Since its discovery, the purpose and age of the Eastern Desert rock art have been the focus of a great deal of debate, and presently there are few definitive answers to the many questions that have been raised. One of the most challenging issues associated with the study of rock art is that it is extremely difficult to securely establish its chronology. Although some frameworks have been developed to allow both inscriptions and painted decoration to be dated on

Figure 7.5. Images of boats carved into exposed rock faces in the Eastern Desert, in this case at Kanais (figure 7.7).

Figure 7.6. Human figures with upraised arms feature prominently in the rock art of the Eastern Desert.

stylistic grounds, the dates that have been suggested tend to be very broad ranges and this method has been found to lack precision. Suggestions that the depth of color of any patina can be used to establish relative dates for individual images must also be treated with care; different layers of exposed strata may develop different patinations, even if inscriptions were added at the same time. Furthermore, the development of patina can be influenced by the presence of moisture, or whether a particular exposure is in direct sunlight or shadow for much of the day. A team led by Dirk Huyge has established dates older than 5000 BC for rock art sites near Edfu in the Nile Valley. These dates were established using carbon dating methods to target specks of organic matter that can be recovered from within the patina (see Further Reading). Although extremely promising, this technology has yet to be used in the Eastern Desert.

One site that may allow rock art to be viewed with a degree of context is Kanais, about 50 kilometers along Wadi Abad, east of Edfu (fig. 7.1). As a small fort and ancient watering station indicate, the well at Kanais was in use during the Greco-Roman period; however, it is clear that the history of the site long predates Roman activity. Nearby is a New Kingdom rock-cut temple, richly adorned with hieroglyphic inscriptions (fig. 7.7). To the east of the temple, the sandstone walls of the wadi and a series of large boulders that fell from the wadi walls many thousands of years ago are decorated with countless images of animals, people, and most notably, a large number of boats (fig. 7.5). While the Roman ceramics and pharaonic hieroglyphs at Kanais can be dated quite readily, the date of the rock art is less certain. By drawing comparisons with predynastic ceramics, however, researchers including Toby Wilkinson (see *Genesis of the Pharaohs* in Further Reading) have suggested that much of the rock art of Kanais was added sometime around 4000 to 3500 BC by visitors from the Nile Valley. Given that most Egyptologists consider that the motifs used to decorate early ceramics had strong religious meaning for the predynastic cultures of the Nile, it also seems reasonable to assume that the boats and other rock art elements that adorn Kanais were similarly imbued with religious significance for the people who inscribed them. At Kanais, therefore, we can perhaps suggest that the rock art was first applied by travelers from the Nile Valley, for whom the presence of a well at this location made it an important focus for early activity in the desert. Perhaps the images of dancing figures evoked the protection of the benevolent Nile gods and the images of river boats brought with them the benign familiarity of the Black Land, now so far away.

There was renewed interest in the rock art of the Eastern Desert in the late twentieth and the early parts of the twenty-first century. A small number of dedicated teams combed the more remote areas beyond Kanais, carefully recording the inscriptions they found in the steep-walled wadis between the Nile Valley and the foothills of the Red Sea Hills. One of the most important contributions of these recent expeditions was the 2004 rediscovery of Site 18 (fig. 7.1). First identified in the 1930s, Site 18 was known to contain

Figure 7.7. The New Kingdom temple at Kanais. Although the temple itself is largely rock-cut, the columns, lintels, and cornice are all added masonry elements.

a number of unusual and potentially important rock art elements, including a number of *serekh*s (fig. 7.8 left). *Serekh*s are rectangular hieroglyphs, often surmounted by a representation of a falcon god, and were used in the Early Dynastic Period to represent the name of the pharaoh. As a symbol of royal authority, *serekh*s were replaced in the Old Kingdom by the more familiar oval cartouche (fig. 7.8 right). Given their clear Early Dynastic origins, the presence of *serekh*s provides a relatively secure early pharaonic date for elements of the Site 18 rock art assemblage. Site 18 is also notable for the variety of animals represented there. These include images of goats and ibex that are a common feature of the rock art of the Eastern Desert, but there are also depictions of more unusual animals such as crocodile, hippopotamus, and ostrich. Many researchers consider animal images to be among the oldest examples of rock art in the Eastern Desert, perhaps having been applied several thousand years before the Early Dynastic *serekh* had been conceived as a symbol of pharaonic power. Such early dates are based on the principle that to sustain animals such as crocodiles and hippopotami, the climate of the area would have been very different from that which existed in the pharaonic era or the latest parts of the Predynastic Period.

Despite the animal images at Site 18 appearing to have been inscribed in a different manner from the *serekh*s, and despite their patina appearing to be significantly deeper in color, they have proved difficult to date. The very early dates that have been suggested by some researchers require us to accept that these animals are literal representations of the environment that existed in the Eastern Desert when these images were created, something that may not necessarily be the case. Particularly significant for this issue is that many of the animal images include unusual features that would not be expected in life, such as the uncharacteristically long tails and exaggerated hooves typically shown on giraffe (fig. 7.9). These observations have led those who argue for a more recent date to suggest the images were drawn from folk memory, long after climatic changes had driven these animals out of the Eastern Desert. We will return to the issue of folk memory when we discuss the Western Desert, in chapters 8 and 9.

The presence of *serekh*s adjacent to images of animals that are considered by many to be somewhat older suggests that the rock art

Figure 7.8. Pharaonic influence in the Eastern Desert. An Early Dynastic *serekh* (left, Site 18—see figure 7.1) and a later cartouche (right, Hammamat quarry) indicating the names of some of Egypt's ancient rulers.

Figure 7.9. Depictions of giraffe and other animals in rock art of the Eastern Desert, at the Jacuzzi (see figure 7.1). Note the unusually long tails and hooves of the giraffe.

of Site 18 may have been applied in a number of distinct phases, just as we saw at Kanais. But it is difficult to apply what we see at Kanais and Site 18 more generally across the Eastern Desert. Although the focus of activity at Kanais was likely to have been the well, like most of the other known rock art sites, there is no well at Site 18. Furthermore, unlike Kanais and Site 18, there are many rock art sites in the Eastern Desert that appear to have been the result of just a single phase of activity.

There is also a general lack of understanding about the purpose that the rock art images may have served. If the animal motifs are indeed some of the earliest elements of the Eastern Desert rock art, perhaps they can be associated with the earliest signs of climate change (chapter 6). Were the images of elephants and giraffes left by the indigenous inhabitants of the Eastern Desert in an attempt to persuade their gods to return the area to a less hostile environment and to encourage a reappearance of these animals? Alternatively, were the early animal motifs offerings to animal deities, drawn by the inhabitants of the region who must at times have felt under threat from visitors from the Nile Valley? As the pharaonic culture developed and trading or mining expeditions were sent out into the desert, were images of protective pharaonic deities, or the names of their powerful ruling pharaohs, left in an effort to placate the fearsome desert gods?

In contrast to the idea that the rock art of the Eastern Desert was motivated by religious or cultural factors, there are reasons to suggest that some elements served more practical purposes. One example of the potential practical role of rock art is an inscription that has been interpreted by some as a map of the local wadi network, although this interpretation has not been universally accepted. In other examples, rock art assemblages include images of camels, which were not known in Egypt until the sixth or seventh century BC, and others include four-wheel-drive vehicles. If we assume that the camel herders or drivers of modern vehicles were simply passing time when they added their inscriptions, could some of the earlier rock art also have been the product of idle hands? Perhaps some of the early animal inscriptions are the work of indigenous goatherds as they sheltered against the shady walls of the wadis, watching their flocks?

I expect that there were a range of motivations that led to the rock art of the Eastern Desert, and that the intentions of the various artists throughout time included both the sacred and the profane. It is my personal view, however, that much of the rock art in the Eastern Desert was left by expeditions from the Nile Valley, exploring the area in search of mineral wealth. As we shall see in chapter 10, the mineral wealth of the Red Sea Hills was exploited not only during the great periods of the pharaonic era such as the New Kingdom, but also during the Predynastic and Early Dynastic periods (fig. 7.10). The rock art of the Eastern Desert may provide evidence for these early mining expeditions, and I am rather drawn to the idea that the images of dancing figures and riverboats were left in an attempt to placate the desert gods and tame the chaos of the Red Land into which these early expeditions ventured. After the unification of the Two Lands, the mining expeditions into the Eastern Desert would have become part of the operation of the

Figure 7.10. Early Dynastic stone pot with gold lid and gold-wire fastenings. It is possible that this gold was quarried during early expeditions into the Eastern Desert.

state, and the addition of royal *serekh*s and cartouches to the wadi walls would have given each expedition royal authority and claimed the protective power of their pharaoh and their gods.

We may never know whether my interpretation is in any way close to reality, and I'm compelled to acknowledge that in focusing solely on the rock art of the Eastern Desert, I have not considered the wider context provided by rock art from other areas of Egypt and further afield. Such uncertainty, however, is perhaps one of the beauties of desert travel. Sitting around a desert campfire at night under a broad canopy of stars, it is fascinating to discuss the rock art of the Eastern Desert and what it may mean. It is perhaps appropriate that in the Red Land—the place of chaos—such discussions can never reach a firm conclusion. Chaos indeed!

CHAPTER 8
THE WESTERN DESERT, PART 1

Lying between the Nile Valley and the Libyan border on the eastern fringes of the Sahara, Egypt's Western Desert is so vast it could swallow up the entire British mainland (fig. 8.1). It is only the fertile oases that break up this inhospitable expanse of rock and sand.

Into the Western Desert

Journeys into such an inhospitable place as the Western Desert require a great deal of planning and can only be undertaken with relevant permits from the Egyptian authorities. A typical expedition will require a fleet of sturdy four-wheel-drive vehicles, highly experienced desert guides and drivers, and an ample supply of food and water. Since the conflict in Libya in 2014, safety concerns have led to very tight restrictions on travel in the Western Desert, but hopefully these will soon be lifted and it will be possible to return to this remarkable wilderness.

As we saw when exploring the Eastern Desert in chapter 7, in most areas west of the Nile, the boundary between *kemet* and *deshret* is represented by the upper walls of the Eonile Canyon. Behind these towering cliffs is an elevated plateau, cut by ancient wadi

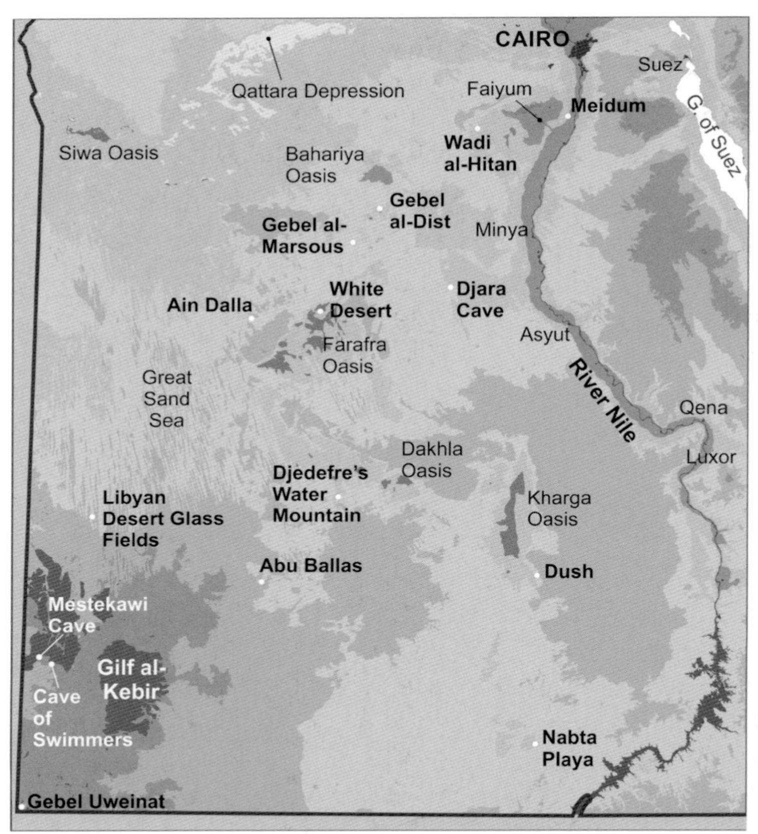

Figure 8.1. Sketch map of the Western Desert showing features referred to in the text.

systems that drained into the Nile Valley and were probably at their most active during the Protonile phase of the river (chapter 6). Of the many Western Desert wadis, a small group opposite the ancient capital of Thebes (modern Luxor) has earned a reputation as one of the most famous archaeological sites in the world. Hidden in these wadis are the burial places of some of the most important individuals from arguably the greatest period in the pharaonic era, the New Kingdom (fig. 1.6). One wadi, the Valley of the Kings, needs no introduction as the burial place of some of ancient Egypt's greatest pharaohs, including the nearly intact tomb of Tutankhamun, discovered there in 1922 by Howard Carter. In chapter 11 we will look at the ancient construction methods used in the Royal Wadis, together with the accident of geology that led to the preservation of Tutankhamun's tomb. Beyond the Valley of the Kings, the Western Desert initially reveals itself as a barren, windswept, and rocky plain with a thin sand cover that extends to the horizon before becoming lost in an indistinct blur of dust and heat haze. As we saw in chapter 5, this haze hides a geological treasure trove of incredible fossils and unusual crystalline hills. The Western Desert, however, holds countless other secrets, including polished rocks from space and the remnants of cultures that inhabited the area in the distant past.

Unlike the Eastern Desert, where you tend to travel along wadi floors, the ancient drainage system of the Western Desert has not had such a clear impact on the landscape; there are fewer deeply incised wadis to follow. Many of the ancient caravan routes that had crossed the exposed surface of the Western Desert have been transformed into modern roads, simply by laying strips of asphalt almost directly over the stony surface. Occasionally, these routes drop sharply as they cross a geological boundary of one sort or another. Many of these descents have little significance; however, about 150 kilometers west of Luxor, the limestone plateau falls away sharply as the road crosses a steep escarpment some 100 meters high. After this initial descent, the route passes over a number of other, less dramatic escarpments, collectively falling some 300 meters before the descent finally comes to a halt. At this point, we are only about 40 meters above sea level, and as the road has followed the contours of this eroded desert landscape, we have traveled back in time. The first major escarpment took us out of the Eocene and into the Paleocene (fig. 2.1) and the subsequent descents took us even further back to Upper Cretaceous sandstones, rocks laid down at a time when the dinosaurs roamed the Earth.

The Drama of the Oases

The road we have taken has followed an ancient route from the Nile Valley to Kharga Oasis, the most southerly of the oases that punctuate the central and northern parts of the Western Desert (fig. 8.1). To avoid the highest cliffs to the north and east of the oasis (fig. 4.3), the road has taken us to the south of Kharga where, despite our 300-meter descent, the escarpment is easier to navigate. At the junction of five desert routes, we come across the first archaeological remains we have seen since leaving the Nile Valley (fig. 8.2). Dush was founded perhaps as far back as the Old Kingdom as a merchant settlement, trading with passing caravans. The town became an important Roman military settlement, and next to the remains of the Roman fort are the ruins of a sandstone temple, once erected in honor of the pharaonic gods Osiris and Isis. Legends say this temple was originally covered in gold, and although such extravagance is unlikely, in 1989 archaeologists did find a hoard of gold jewelry of the highest craftsmanship among the ruins of Dush. Despite this great wealth, the merchants of Dush could do nothing to prevent their wells from drying out, and during the fourth or possibly fifth century AD, the town was abandoned. The fate of Dush underlines the single most important fact in the desert: life depends entirely on the availability of water.

Excluding the region of Faiyum, there are five major oases in the Western Desert, from Kharga in the south to Siwa in the north (fig. 8.1). Each of these fertile oases lies at the base of a depression, a natural low-lying area that has been eroded into the surface of the desert, with high escarpments defining their northern and, in some cases, eastern extents (fig. 4.3). The oasis depressions are large (Bahariya is about 90 kilometers from north to south) and the floors of the oases generally lie only 100 meters or so above sea level at their lowest points. To the oases we can add the Qattara Depression (to the northeast of Siwa, fig. 8.1), which is about 160 kilometers wide and reaches about 130 meters below sea level, making it one of the lowest-lying points in Africa. Both the depressions in which the verdant oases lie and the barren and inhospitable Qattara Depression all appear to have been formed in much the same way, and yet their environments are so very different. The reason for this stark contrast is that in the oases, erosion has reached the Nubian Sandstone aquifer, and it is the ready availability of precious groundwater that has transformed these low-lying areas into fertile and habitable regions in the midst of the desert. As discussed in chapter 4, there is concern that

Figure 8.2. Remains at Dush.

over-exploitation of the Nubian Sandstone aquifer is lowering the groundwater table, leading to what may be irreversible changes in the hydrogeology of the Western Desert and the oases. Together with natural changes in groundwater regime, such as those that caused the wells at Dush to dry up in antiquity, the impact of modern groundwater extraction on the Nubian Sandstone aquifer represents one of the major challenges facing modern Egypt.

There has been much debate about how the depressions of the Western Desert formed. Long-standing theories suggest that they are the result of erosion caused by wind, but it is difficult to explain how such a relatively inefficient process could create such large and relatively deep basins in the desert's surface. Among the alternative theories that have been put forward is the suggestion that the depressions are the result of collapse of subsurface cave systems running through the soluble limestone strata with which the depressions are generally associated. Although caves are known in the Western Desert, they are generally not of a scale that matches the oasis depressions. Perhaps the most spectacular of these caves is the relatively inaccessible stalactite- and stalagmite-infested cave at Djara, over 100 kilometers southeast of Bahariya (fig. 8.1). Smaller and perhaps less spectacular is the more readily accessible Obayid cave in the limestone escarpment to the north of the Farafra Oasis. This small cave is perched at a strategic position midway up the escarpment, and the cave's mouth provides a vantage point that dominates the surrounding desert. The cave was clearly occupied in antiquity, in all probability long before Egypt's pharaonic cultures had emerged in the Nile Valley. The cave's walls and ceiling are decorated with shallow engravings of what may be antelope with round distended bodies, together with arrangements of different-sized circular depressions that resemble stylized bear paws. Another remarkable element of the rock art in Obayid cave, and one that we will see again as we travel deeper into the Western Desert, is the red ochre images of the outline of human hands, which were created by ancient desert dwellers holding their hands against the rock face and spraying over them with pigment. These very personal images force us to think about the individuals who lived in this area many thousands of years ago.

If the formation of depressions in the Western Desert was the result of the collapse of cave systems that had developed in the limestone strata, what caused these caves to collapse? We may have already identified one possible answer to this question when we discussed the Gilf River in chapter 5. Although the Gilf River system (fig. 5.2) is now hidden beneath the thick, drifting sands of the Great Sand Sea, when the river was active it would have eroded the surface of what is now the Western Desert, simultaneously advancing its headwaters to form high escarpments such as those that surround many of the oases and cutting down into the limestones that blanketed the area (plate 3). In areas of cave systems, this down-cutting may have reduced the thickness of the overlying limestones to such an extent that the caves collapsed, leading to the formation of the depressions that are now occupied by the oases and the Qattara Depression. Under this scenario, the escarpments that surround the oases may represent the furthest points of advance of the Gilf River system, the points reached by the river's headwaters before the climate of the region changed and this great river lost its dominant role in the evolution of the Western Desert landscape.

The Great Sand Sea

The Great Sand Sea is the name given to a whirling sea of shifting sand that covers an area of about 100,000 square kilometers in the central Western Desert (fig. 8.1). The Great Sand Sea extends for about 650 kilometers from north to south and some 300 kilometers from east to west and was named by the first known Western explorer, Gerhard Rohlfs, who crossed the area in 1869. As we have already seen, the source of these vast quantities of sand was the Gilf River system, which eroded the limestone and sandstone terrain across a vast area west of today's Nile Valley and transported the eroded sediment downstream to form an extensive delta that lay along the ancient coast. As regional earth movements forced this coast to retreat northward, these accumulations of sand were exposed to a newly dominant north wind that, from about 10 million years ago, began to blow these deposits back on land, eventually burying the valleys of the Gilf River system and covering them with a vast carpet of sand. Today, the Great Sand Sea is dominated by thousands of massive, elongated dunes that are aligned to the northerly wind and are steadily and progressively moving southward (fig. 8.3). On the ground, these swarms of sand dunes resemble a huge ocean swell, each dune a breaking wave that can extend for hundreds of kilometers from north to south and are many tens, or possibly hundreds, of meters high. The only way to cross the Great Sand Sea is to follow a widely meandering route, seeking out the points where the dunes are at their lowest (plate 10). Even with these precautions, however, a journey across the Great Sand Sea becomes an incredible roller-coaster ride that, despite the vast experience of desert drivers, will occasionally result in vehicles having to be dug out of the soft sand in which they have become trapped.

Not only is the Great Sand Sea a dramatic landscape by virtue of the enormous dunes and the subtle, almost magical effect that the wind has on their ever-changing appearance, there are other features that add to the mystery of traveling through what is a very alien world. During the Second World War, this area of North Africa was the scene of a great deal of clandestine military activity, with both the Allied and Axis forces using the desert

Figure 8.3. Dunes in the Great Sand Sea.

to cloak their long-range reconnaissance missions. The difficulty in using vehicles in this environment is brought home by the remains of abandoned trucks, former fuel dumps, and other military paraphernalia that are occasionally found in the desert (fig. 8.4). Although they have lain abandoned for over 70 years, these vehicles are in remarkably good condition. The metal has rusted, but this is a strange type of rust, which in the absence of any significant moisture has resulted in the exposed metal taking on a smooth, deep chocolate brown patina. Both the corroded metal and the rubber of the tires have been polished by the action of millions of sand grains as they brush past, carried by the wind.

Even in such an apparently unpromising place as the Great Sand Sea, there are archaeological wonders to be found. But these are not from the pharaonic era; they are much older, giving us a glimpse of some of mankind's earliest activity in Egypt. For a long time, Stone Age archaeology in Egypt was overshadowed by the pharaonic era; however, as discussed by Justin Maxwell Heath (see Further Reading), a great deal of research is currently underway to develop a more thorough understanding of mankind's earliest activity in Egypt. Despite this ongoing research, the early Stone Age (more correctly referred to as the Paleolithic Period) remains poorly understood. The oldest phase, the Lower Paleolithic, ended about 200,000 years ago and is characterized by distinctive pear-shaped Acheulean hand axes, examples of which have been found in the Great Sand Sea, the oases, and the Nile Valley. Hand axes found near the Qena Bend are thought to be as much as 300,000 years old. Limited excavations of Acheulean sites in the Kharga and Dakhla oases have yielded later examples (about 250,000 years old), and although other Acheulean sites are suspected in areas outside the oases, systematic investigations have yet to be undertaken. Most remarkably, rather than being the tools of modern man, these hand axes are thought to have been made by *Homo erectus*, a highly successful but now extinct human species that may represent one of modern man's ancestors.

In addition to these remarkably early man-made artifacts, there are many natural wonders hidden among the dunes of the Great Sand Sea. In places across the higher dunes, thin, spidery veins can be seen radiating across the sand. Closer inspection reveals these veins to be surprisingly lightweight, glassy tubes perhaps no more than 7 or 8 millimeters across with thick, irregular walls (fig. 8.5). They are so fragile that they readily break under the action of the wind. Fulgurites, as these strange veins are called, are the result of lightning in the desert. With no trees or other tall natural features, lightning tends to strike the highest dunes, and the intense electrical energy of the lightning bolt locally melts the sand grains. The veins we see in the desert mark out the path that the lightning bolt has taken as it ran to earth, locally melting the sand and turning it to a form of natural glass.

In one relatively small area in the south of the Great Sand Sea, the shifting sands reveal yet another mystery of the Western Desert. Translucent yellow or green in color and with an unusual soapy texture, fragments of silica glass litter the surface of the desert between

Figure 8.4. An abandoned truck from the Second World War, slowly rusting in the arid conditions of the Great Sand Sea.

the high dunes (fig. 8.1 and plate 17). The origins of Egypt's silica glass, or Libyan Desert glass (LDG) as it is often called, has been hotly debated for many years, and as discussed by John Saul, there is still little agreement among specialists (see Further Reading). About the only feature of LDG for which there is agreement is its age, which, at a little short of 30 million years, suggests that it formed at a time when the landscape of this part of Egypt was characterized by flowing rivers and dense forests (chapter 5).

From a global perspective, silica glass is not unusual, having been found in association with meteorite impact sites across the Earth. When meteorites strike, the extremely high energies involved rapidly melt the impacted soils and rocks, ejecting the molten material away from the impact site. As this ejecta travels through the air, it cools to form lumps of natural glass that are rich in silica, one of the Earth's most abundant minerals. One of the unusual features of LDG, however, is that it is *extremely* rich in silica, with concentrations generally greater than 96 percent. Some researchers have suggested that the abundance of silica in

Figure 8.5. "Vitrified lightning"—fulgurites in the Great Sand Sea, formed when lightning strikes the top of dunes. These examples are less than 10mm across.

LDG is consistent with a meteorite impact in areas of Nubian Sandstone (plate 3); however, despite examination of a number of possible impact features in the region, no suitable site has been identified. It is possible that the point of impact may have been a considerable distance from the silica glass fields, with the smoothed appearance of many of the LDG fragments possibly due to transport by water (plate 17). The possibility of transport from remote areas may also be consistent with other research, which has identified trace components of LDG that are more typical of distant outcrops of igneous rocks. One difficulty with any theory that involves transport over significant distances, however, is that this is likely to have distributed the LDG fragments across a relatively large area, whereas LDG is present only in a fairly restricted area of the Great Sand Sea.

The difficulties that have been encountered when trying to identify evidence for the impact origin of LDG have led a number of researchers to seek other explanations for its formation. In 1996, an unusual fragment of black stone was discovered in the silica glass fields of the Great Sand Sea. This stone was given the name Hypatia and detailed analysis suggests that it might be something very rare indeed. Hypatia was found to contain ratios of oxygen and carbon that are not generally found in terrestrial rocks; however, the gases trapped within the specimen did not resemble those found in conventional meteorites. By comparison with data from other extraterrestrial bodies, Jan Kramers and colleagues have concluded that Hypatia may in fact be the first fragment of a comet ever found on Earth (see Further Reading). It has been suggested that the immense energy released from a comet exploding in the lower atmosphere over the Western Desert could readily explain the formation of LDG without the need for an accompanying impact crater. Although further proof is needed, the discovery of Hypatia in the same relatively small area of the Earth's surface as LDG suggests that these materials may be linked, pointing to a possible cometary origin for Egypt's silica glass.

A number of pre-pharaonic artifacts that were fashioned from fragments of LDG have been discovered in the Western Desert, including a hand axe that is largely Acheulian in character and most probably dates from the Lower Paleolithic Period. These very early artifacts date from a time when the deserts in Egypt were less arid and would have been occupied (if only sparsely) by small populations of hunter-gatherers or other nomadic or semi-nomadic people. Perhaps the biggest surprise of recent years was the discovery that the Egyptians of the New Kingdom also knew of LDG and perhaps even valued it. During the New Kingdom, the Western Desert would have been as dry and inhospitable as it is today, so it is remarkable to find a large piece of polished silica glass, carved into the shape of a scarab, as the centerpiece of a pectoral discovered in the burial of Tutankhamun (plate 18). The pectoral was an important element of pharaonic burial regalia, laid on the chest of the pharaoh's mummy. For many years, the strange green stone at the heart of this piece had been misidentified as chalcedony, a relatively common stone far less valuable than the gold into which the scarab had been set, and perhaps not in keeping with the important role this pectoral played in the burial rites of an Egyptian pharaoh. The recent reclassification of this scarab as a specimen of worked LDG indicates that the ancient Egyptians were familiar with the Western Desert and its resources. It seems unlikely, however, that they could have appreciated the unusual origins or the rarity of LDG.

The presence of a piece of LDG in the Tutankhamun pectoral leads to some very important questions: What did the ancient Egyptians know of the expanses of the Western Desert beyond the oases? Were they motivated to explore the inhospitable areas beyond the fringes of the pharaonic empire, and if so, how could they have known that there was anything out there worth the effort? Having thought about this question a great deal, my view is that we are somewhat blinkered by our modern perspective. Under the conditions that exist today, desert travel is a difficult undertaking and when we are standing in the Nile Valley or in one of the oases, there is little obvious benefit to be gained by setting out west into the depths of the Red Land. Unlike the Eastern Desert and the Red Sea beyond, the Western Desert has no obvious trading destination, nor any significant concentrations of mineral wealth. I have deliberately chosen to refer to the Western Desert as the Red Land here because, as we have seen, this ancient concept speaks of the desert as a place of chaos and danger. Our modern understanding of ancient attitudes to the Red Land is that it was a place to be shunned in favor of a life in the fertile Nile Valley. Despite the trepidation that the deserts must have instilled in the minds of the ancient Egyptians, however, it is increasingly clear that if we are to understand the archaeology of ancient Egypt, we need to more fully appreciate the extent to which the Red Land was explored by the people of the Nile Valley.

It is my personal view that the key to deepening our understanding of these issues (and we are only *beginning* to get a clearer picture of ancient activity in the Western Desert) is to shift our perspective. As we have seen, from perhaps 10,000 years ago the Western Desert would have experienced a period of climatic transition, with the onset of summer monsoon rains making the early Holocene climate less arid. Archaeological evidence, mainly based on worked stone implements, indicates that at this time, the far reaches of the Western Desert were occupied by nomadic hunter-gatherers, people with cultural origins that were quite distinct from those of the inhabitants of the Nile Valley. By about 8,000 years ago, the continuation of the monsoon rains had made the Western Desert more widely inhabited. Although a nomadic way of life would have still been important, the height of the Holocene Wet Phase can be identified in the archaeological record by settlements that were occupied for significant parts of the year, particularly in the oases, which are likely to have covered a greater area than they do today. From about 6,000 years ago, the regional climate began to dry out again and the nomadic way of life would have become progressively more difficult, drawing most of the inhabitants toward the oases or the Nile Valley. As we will see, however, there is evidence to suggest that more favorable conditions persisted in a number of limited areas of the Western Desert for some time after the Holocene Wet Phase ended. It is these isolated areas and the people that occupied them that are perhaps the origins of legends such as that of Zerzura, a fabled oasis in the Western Desert that has now become lost in the mists of time.

The Changing Cultures of the Western Desert During the Later Holocene

The period of increasing aridity that marks the end of the Holocene Wet Phase began about 6,000 years ago and appears to have ended in the late Old Kingdom (about 4,000 years ago; see fig. 1.6). As discussed at a recent conference on the archaeology of the Western Desert (edited by Gillian Bowen and Colin Hope; see Further Reading), excavations, particularly in the Dakhla and Kharga oases, are revealing evidence for a number of distinct indigenous cultural groups that occupied the Western Desert during this period. Thanks to these recent excavations, we are able to shed some light on the response of these cultures to the changing environment in which they lived. The Bashendi people occupied the Western Desert at a time roughly comparable with the early Predynastic Period in the Nile Valley, and can be recognized by their stone tool assemblage mixed with small amounts of distinct pottery. Although initially the Bashendi appear to have been nomadic, during the height of the Holocene Wet Phase they adopted an increasingly sedentary way of life, before the onset of arid conditions led them to return once more to a predominantly nomadic lifestyle. Somewhat later comes the Sheikh Muftah culture, known principally from finds in and around the Dakhla Oasis. Although they seem to have settled for certain parts of the year, the Sheikh Muftah people were also predominantly nomadic and their distinct culture appears to have survived until the end of the Old Kingdom, at which point they disappear from the archaeological record. It should not be assumed that the disappearance of the Sheikh Muftah people is an indication that they were suddenly "swallowed up" by their pharaonic neighbors. The archaeological record suggests that the two cultures enjoyed a lengthy period of successful cross-fertilization of ideas before the cultural traits of the Sheikh Muftah people became subsumed into the increasingly successful pharaonic culture that was highly active in the oases by the late Old Kingdom.

Set against the backdrop of a changing climate, the evidence for activity by pharaonic Egyptians in the Western Desert is also extremely interesting. Although sparse, evidence for Early Dynastic activity in the Western Desert is significant and suggests that the pharaonic Egyptians of this period were becoming increasingly active in areas outside the Nile Valley. Researchers consider the most likely reason for Early Dynastic activity in the Western Desert was hunting or trade, and there is some evidence that at that time, the pharaonic Egyptians may have worked alongside guides from the Sheikh Muftah culture. It was not until the end of the Early Dynastic Period, and certainly by the onset of the Old Kingdom, that there is evidence of pharaonic settlements in the oases. The exact purpose of the early Old Kingdom interest in Dakhla has yet to be established; however, this does not appear to be the activity of an occupying force, further suggesting that the motivation lay in hunting or trade. At this stage, the pharaonic occupants of the oases maintained very close ties with the Nile Valley, with an absence of pharaonic burials from this period suggesting that the deceased may have been returned to the Nile Valley for interment. After a period of little apparent interest in the oases in the middle of the Old Kingdom, the late Old Kingdom saw a relatively sudden expansion of pharaonic settlements, with appointed governors ruling on behalf of the pharaoh.

Well-established oasis settlements such as Balat in southwest Dakhla underwent significant growth as official political residences, civilian houses, workshops, and granaries were constructed, together with the establishment of necropolises that, although bearing provincial traits, were essentially pharaonic in character.

It is interesting to speculate on the reasons for the increased late Old Kingdom interest in the Western Desert, and although it would be easy to identify this as a period of empire building, a lack of evidence for widespread conflict in the oases at this time suggests otherwise. As we have already seen, the period of increasing aridity from 6,000 years ago may have led to episodes of sustained drought in the Nile Valley during the late Old Kingdom. When viewed in this context, I wonder whether the renewed interest in the oases at that time was brought about by a need to increase pharaonic Egypt's access to readily cultivable land, to help feed the population of the Nile Valley.

Just as the Sheikh Muftah culture appears to have integrated with pharaonic Egyptians that occupied the Western Desert oases, I also wonder whether similar processes may have taken place elsewhere in Egypt. As indigenous populations migrated out of the increasingly arid desert, did they become subsumed into the host population of the Nile Valley and the oases? Even if this was the case, it may be a mistake to assume that these indigenous cultures were lost completely; aspects of their customs and beliefs may have been adopted by the pharaonic populations into which they were absorbed. In this way, the pharaonic culture may have been influenced by the culture of people that had once inhabited large parts of northeast Africa. This cultural assimilation may have also led to the preservation of certain aspects of folk memory within the population of pharaonic Egypt, allowing certain details of the landscape, or of the people and places of the Western Desert, to survive. In this way, it is possible that a vague understanding of the areas beyond the oases was preserved, and with it, the knowledge that the deserts could be crossed. In contrast to our modern view that the Western Desert is a vast, lifeless, and largely uncharted wilderness, for the people of Early Dynastic and Old Kingdom Egypt at least, the Western Desert may have been a far less forbidding place, a place full of distant memories and possible trading opportunities.

CHAPTER 9
THE WESTERN DESERT, PART 2

We ended the last chapter with discussions of folk memory and the possibility of distant recollections of long-lost cultures far across the Western Desert. Even with the best modern equipment, expeditions beyond the oases are serious undertakings that require the utmost planning and preparation. How were the pharaonic Egyptians able to explore this inhospitable wilderness?

Desert Roads

In chapter 6, we discussed the important role that communication along the River Nile played and the importance of river travel in allowing pharaonic Egypt to function as a single nation-state. In addition to river travel, a number of key overland routes have long been recognized to the west of the Nile, including routes that bypassed the lengthy sailing around the Qena Bend (fig. 1.1) and routes that connected the Nile Valley to the oases. Recent research, however, has shown that the ancient road network of the Western Desert was far more complex than previously believed, with trading routes running south and west from the oases, possibly into the very heart of Africa (see Frank Förster and Heiko Riemer in Further Reading). The rapid growth of Balat toward the end of the Old Kingdom (chapter 8) may have been connected with its role as the northeastern terminus of one of these cross-desert routes, known today as the Abu Ballas Trail.

Out beyond the oases and away from the huge dunes of the Great Sand Sea, the flat, sandy wastes of the Western Desert are interrupted occasionally by isolated groups of sandstone hills that even today serve as way-points, prominent features that aid navigation across this otherwise featureless terrain. In ancient times a number of these hills, including Abu Ballas (fig. 9.1) and the oddly named Djedefre's Water Mountain (fig. 8.1), were used not simply for navigation but also fulfilled a far more vital role. Abu Ballas means "Father of Pots," and when rediscovered in the early part of the twentieth century, the foot of the hill was surrounded by hundreds of ancient ceramic jars of a type well known from the Nile Valley and the oases. Archaeological work at the site has been hindered because over the intervening years, many of these pots have been illicitly taken away by visitors to this remote spot, and today only fragments survive. The name given to Djedefre's Water Mountain comes from a hieroglyphic inscription that was found there, which includes the cartouche of Djedefre, a fairly short-lived monarch of the Old Kingdom (fig. 11.3). In contrast to Abu Ballas, this site has only recently been discovered, and its relatively undisturbed state has allowed researchers to develop a much greater understanding of the role these sites played in supporting ancient desert travel. The numerous pots found at these sites were used for storing a range of basic goods such as water and grain and were associated with encampments where small fires were used not only for warmth, but also for the preparation of bread and other basic foodstuffs. Strung out along the key cross-desert routes, these sites provided ancient travelers with the opportunity to break their journey, rest or even exchange their animals, restock their water, and take on fresh provisions before venturing forth on the next leg of their desert trek.

Given the apparent size and complexity of these support operations, it seems clear that the desert road network and the expeditions that ventured along it were maintained and operated only with the full support and resources of the pharaonic state, once more pointing to trade as their most likely motivation. Together with the evidence

Figure 9.1. An isolated hill in the Western Desert, one of a group of hills that includes Abu Ballas. The lower slopes of Abu Ballas can be seen in the right of the photograph. Prominent groups of hills such as this may have helped the ancient Egyptians navigate in the desert.

from sites such as Balat, it is clear that these desert resupply points were established relatively early in the pharaonic era, and although evidence has been found for a brief period of renewed interest in the New Kingdom, it is becoming increasingly apparent that the greatest period of desert exploration took place in the Old Kingdom, possibly before much of the folk memory associated with the less arid conditions of the Holocene Wet Phase had faded (chapter 8). If Egyptians of the Old Kingdom were prepared to launch trading expeditions across the barren landscapes of the Western Desert, it raises questions about the cultures they were trading with.

The Gilf al-Kebir

One possible answer to this last question may be found in the deep southwest of Egypt, in an area dominated by a mass of uplifted and durable sandstones and siltstones that were laid down in the early part of the Mesozoic Era, some 200 million years ago (fig. 2.1). Time and nature have been unable to erode these rocks as rapidly as the younger Nubian Sandstones that surround them, resulting in an impregnable 300-meter-high plateau the size of Wales or Switzerland, known as the Gilf al-Kebir (fig. 9.2). Gilf al-Kebir can be translated as "The Great Barrier," and it is clear why this incredible landscape feature was given that name. Sheer cliff-like walls extend for almost 500 kilometers around the eastern, western, and southern flanks of the towering Gilf al-Kebir plateau, and it is only from the north that it presents a less dramatic façade, with the northward-dipping rocks being slowly drowned by vast waves of sand, the relentlessly advancing dunes of the Great Sand Sea.

One of the most prominent modern explorers of the Gilf al-Kebir was Brigadier Ralph Bagnold, who mounted a number of expeditions in the 1930s, publishing a series of reports and progressively updating the existing maps of the Western Desert as his team ventured south and west, out of the oases. Given the Gilf's impregnable cliffs, initially Bagnold's team could only explore the base of the plateau, painstakingly probing every significant wadi for a route to the plateau surface. As Bagnold's published maps indicate, many of these wadis were found to be impassable, largely due to sand dunes and sand drifts that formed barriers across them, and it took until 1939 before the great breakthrough came. The top of the Gilf al-Kebir plateau is so otherworldly that it has been considered for training astronauts for manned missions to Mars.

Figure 9.2. The towering cliffs of the Gilf al-Kebir, viewed from the southeast.

Figure 9.3. Looking west into Libya from the top of the Gilf al-Kebir. A wadi runs away from the foot of the cliffs (arrowed).

In some instances, the generally flat upper surface of the plateau is interrupted by mountainous ridges that project upward like ruined battlements of a lost clan of giants, but not even these provide any indication of distance or scale across this vast, elevated landscape. Bagnold's expedition found ancient tracks crossing the Gilf and the remains of stone circles on the top of the plateau, which are possibly signs of ancient settlement. From the information Bagnold published, however, we have no indication of the age of these features. Remarkably, for an area on the fringes of the Sahara, there are frequent references to water in Bagnold's reports and maps. He published photographs that appear to show recently active wadis, and even today, incompletely filled wadis that may have drained

recent rainfall can be seen radiating away from the base of the Gilf (fig. 9.3). Bagnold identified vegetation growing in a number of the wadis he explored, presumably in areas that had retained a little moisture from the most recent rains. At one location at the foot of the cliffs in the northwest of the Gilf, Bagnold identified a "mud-pan with vegetation and implements," which suggests that this area may have had an even greater tendency to retain rainwater, and had therefore been a focus for ancient activity. Again, Bagnold's reports give us little further detail to indicate the age of the activities his team had identified.

In the decades since Bagnold's work, researchers have sought to establish a chronological framework for the climate and occupation of the Gilf al-Kebir. In a remarkable twist of fate, one of the keys to unlocking this chronology may be a sand drift that Bagnold had actually identified on his 1939 map, blocking Wadi al-Bakht in the east of the plateau. Recent studies by Heiko Riemer and his colleagues have identified that rainwater flowing off the upland areas of the Gilf had formed a seasonal lake (or playa) behind this sand drift (see Further Reading). Analysis of the layers of sediment making up the playa deposits, along with carbon dating of material from organic-rich layers, has permitted the climatic history of the Gilf to be reconstructed for a vitally important period associated with the Holocene Wet Phase, from about 8500 BC to 3500 BC. The evidence from Wadi al-Bakht largely confirms what had been indicated by other regional climate studies: that for much of this period, this part of the Western Desert experienced monsoon conditions dominated by heavy summer rains. The layers of organic-rich sediment in the Wadi al-Bakht deposits indicate that these summer rains were able to support relatively extensive seasonal vegetation. Interspersed with the organic layers, however, are layers of wind-blown sand, demonstrating that between the monsoon rains, the area experienced prolonged dry periods. The evidence from Wadi al-Bakht also confirms that monsoon conditions in Egypt ended at about 4000 BC (about 6,000 years ago). However, unlike the majority of the Western Desert, which became progressively more arid from this time, it appears that the Gilf al-Kebir continued to enjoy relatively high annual rainfall totals for several centuries after the monsoon conditions had ceased. The evidence from Wadi al-Bakht indicates that these post-monsoon rains were more evenly distributed throughout the year and capable of sustaining longer-lived vegetation such as grasses. Evidence for occupation at the surface of the playa deposits suggests that the more sustained rainfall conditions lasted until about 3500 BC, by which time the playa had dried out completely. This date of about 3500 BC is remarkably close to the start of the pharaonic period, which is generally accepted to be about 3000 BC (fig. 1.6).

Although unquestionably an arid landscape at present, the evidence from both Bagnold's expeditions and more recent studies makes it clear that rainfall over the Gilf al-Kebir plateau persisted until the very dawn of pharaonic Egypt. Along with other nearby upland areas such as Gebel Uweinat, some 150 kilometers to the southwest (fig. 8.1), the Gilf al-Kebir was probably among the last areas in this part of the eastern Sahara to have experienced significant annual rainfall and, as such, may have been occupied for several hundred years after the surrounding areas of the desert had become uninhabitable. Was it the people who continued to inhabit places like Gilf al-Kebir and Gebel Uweinat that the Early Dynastic and Old Kingdom expeditions from Balat were seeking to trade with? Was it these cultures that spawned legends of lost oases that were thought to exist deep in the Western Desert? It was the search for the legendary Zerzura that motivated the Hungarian count László Almásy to embark on a series of expeditions into the Western Desert in the 1930s, expeditions that provided the inspiration for the novel and film *The English Patient*, and it was during this great period of exploration that some of the desert's best-kept secrets were revealed to the modern world.

The Cave of Swimmers

Located in Wadi Sura at the southwest of the Gilf al-Kebir, the Cave of Swimmers is not actually a cave, but a hollow at the base of a rocky pedestal that stands a little to the south of the main Gilf plateau (figs. 8.1 and 9.4). When the Cave of Swimmers was first discovered by explorers in the early part of the twentieth century, the distinctive form of painted rock art that characterizes the site

Figure 9.4. The Cave of Swimmers: a natural hollow at the base of this sandstone outlier at the southern edge of the Gilf al-Kebir.

Figure 9.5. Rock art in the Cave of Swimmers. The figures were painted onto the exposed sandstone.

was completely unknown. As exploration of the Western Desert has continued, however, the number of identified painted rock art sites has increased substantially, with in excess of 1,000 sites having been identified in areas including both the Gilf al-Kebir and the uplands of Gebel Uweinat. Despite the ever-increasing number of painted rock art sites in the region, however, the rock art of Wadi Sura still retains a unique character that has not been seen elsewhere.

Figure 9.6. Depictions of people in the Cave of Swimmers. Are these people playing a clapping game (left) and arguing or fighting (right)?

The people who created the rock art of the Cave of Swimmers used pigments they made from mineral-rich Paleozoic strata that outcrop along Wadi Sura and further to the west of the Gilf (the limited areas colored gray in plate 3). The painted images of people and animals are generally rendered using a deep red-brown pigment; however, there are examples where yellow paint has been used, and others where images of cattle have been painted in red and white. In many respects the paintings are executed simply, with each image being represented by the fewest possible strokes, and yet despite this simplicity, they seem so full of life. In addition to swimmers (fig. 9.5), figures seem to be clapping or playing games (fig. 9.6 left), fighting, dancing, and gesturing wildly to one another (fig. 9.6 right).

Mestekawi Cave

Some 15 kilometers to the northwest of the Cave of Swimmers there is a more recent and in many ways more spectacular discovery: another rock shelter known as the Mestekawi Cave, or Cave of Beasts (fig. 9.7). Unlike the Cave of Swimmers, Mestekawi Cave is high above the floor of the desert and it is necessary to climb up a loose, sandy bank to reach it. It is impossible for me to give an impression of the impact that the decoration of Mestekawi Cave has when you first set eyes on it. In terms of the number of individually painted figures, the rock art of Mestekawi dwarfs that of the Cave of Swimmers. Like the smaller site, there appears to be no

Figure 9.7. A small sample of the rock art of Mestakawi Cave.

Figure 9.8. Depictions of people at Mestekawi Cave. A group of tall, muscular people (left) and broad-shouldered individuals (right).

overall composition to the thousands of painted figures of people and animals that are rendered in red, yellow, or occasionally white pigment. There are swimmers at Mestekawi, but these are relatively few in number. In many cases, groups of people appear to be dancing, while others are deep in discussion, possibly over matters of extreme importance. Some of the people are shown as tall and muscular (fig. 9.8 left), others are stocky and broad-shouldered (fig. 9.8 right), raising questions about whether the depiction of people with different physiologies is deliberate, perhaps a reflection of different cultures that met at the Gilf.

The human hand appears to have been especially important at Mestekawi, with a number of individual figures appearing to have been portrayed with exaggerated or emphasized hands (fig. 9.9). In contrast to the relatively isolated examples at the Cave of Swimmers and Obayid Cave (chapter 8), the negative imprints of hundreds of hands adorn the rocky walls at Mestekawi (plate 19). Each hand is a reminder of the humanity of these people: individuals with a desire to leave a record of their visit to what must have been a very important place in their spiritual lives. Amid all of these painted, outstretched, seemingly waving hands, the clearest example of this individuality is provided by a single clenched fist that was captured as the pigments were blown onto the rough sandstone walls of the rock shelter (plate 19, top left). We can assume that this person was shaking their fist at the world they inhabited, but it is fun to speculate whether they could have conceived of a distant future in which visitors to this remote spot might wonder why this individual felt so compelled to express themselves in a manner so strikingly different from their peers.

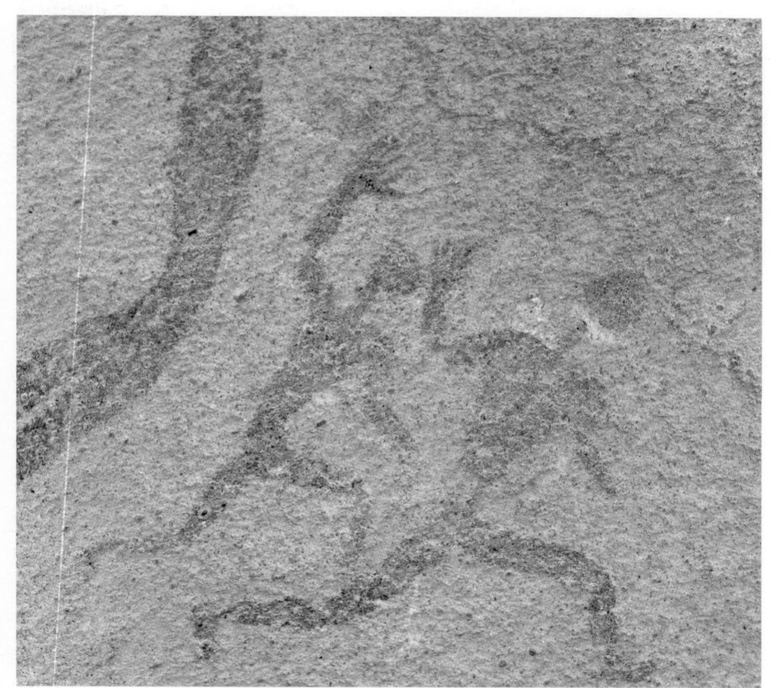

Figure 9.9. Mestekawi Cave. Muscular figures with highly exaggerated right hands, but no evidence of a left hand.

In addition to the images of people, Mestekawi preserves a greater number of animal depictions than the Cave of Swimmers, with the animals rendered in a number of different ways. In one case, an unpainted line of what appear to be antelope, with prominent horns and strangely elongated hooves, have been carved into the sandstone surface. In most other cases, however, animals are painted rather than carved, occasionally using two colors. There is at least one giraffe, though this is damaged and has lost its head, together with a number of other "beasts" that cannot be identified because they are also headless. In these cases, however, the absence of a head appears to be deliberate, with all the evidence suggesting that these beasts were originally painted this way. From the body shapes, some of these headless animals may be lion, buffalo, or rhinoceros, but it is hard to be certain.

The painted images at both the Cave of Swimmers and Mestekawi Cave have faded in time. In addition to this fading, the sandstone surface that served as the canvas for this rock art is deteriorating naturally. Although the images are sheltered to some extent within their shallow hollows, the extreme temperature differences experienced in this desert environment, with hot days and cold nights, coupled with the effects of any moisture that may condense on the cold surface, have led to the exposed sandstone suffering from exfoliation, in which thin slivers of rock peel away from the surface beneath. These natural processes have been exacerbated by attempts to deliberately (and illegally) remove some of the decoration. Collectively, this deterioration represents a great threat to the paintings in Wadi Sura.

Interpreting the Wadi Sura Rock Art

Thanks to the painstaking investigations that have been undertaken in the areas of the Gilf al-Kebir and Gebel Uweinat in recent years, it has been possible to develop some understanding of the Wadi Sura rock art. Although it is clear that many of the human and animal figures in the Cave of Swimmers or at Mestekawi Cave are gathered in groups, there is nothing to indicate that the decoration of these rock shelters was undertaken as a single, overall composition. As discussed by Heiko Riemer and his colleagues, detailed surveys have established that, just as we saw at some locations in the Eastern Desert (chapter 7), the rock art of Wadi Sura may have been applied in a number of phases. The negative images of hands appear to be the oldest element of the rock art corpus and, as such, are probably least understood. There are examples in which the hand prints appear to have been overpainted by humanlike figures and headless beasts. In many cases, the contrast in the depth of color between the two sets of images suggests that these paintings are widely separated in time, though the evidence for this is far from conclusive. The less faded colors of the overpainted figures suggest that they represent a secondary phase of occupation that, thanks to the analysis of the playa deposits at Wadi al-Bakht, can perhaps be dated to the period from about 6500 BC to 4500 BC. Stylistically distinct images of cattle and cattle herding have been identified as the youngest of the rock art elements, associated with a pastoralist culture that took advantage of the more evenly distributed rainfall that characterized the Gilf al-Kebir after the monsoon rains had ended, sometime after 4500 BC.

Although there is currently no academic consensus on this, it does seem reasonable to assume that some aspects of the religion and culture of pharaonic Egypt had their origins in other traditions. With this in mind, it is interesting to speculate whether the relatively small number of yellow painted figures at both the Cave of Swimmers and Mestekawi were intended to represent deities, just as the pharaonic Egyptians considered gold to be the flesh of the gods. Similarly, researchers have suggested that the headless animals painted at Wadi Sura may be examples of symbolic magic, similar to that used in pharaonic Egypt. In many instances of pharaonic art, dangerous animals were symbolically "controlled" by having them shown as dismembered, often by a "magical" knife. Did the ancient cultures of Wadi Sura use similar imagery to limit the power of dangerous animals by representing them without a head and thus depriving them of their source of power: their jaws or horns? An interesting theory put forward recently by Miroslav Bárta (see Further Reading) is based on an interpretation of a group of figures at Mestekawi. Unusually, the headless central figure in this group is painted white and is shown in such a peculiar posture that it is difficult to tell whether it is human or animal. Bárta considers that the manner in which the body of this figure is arched and the way the limbs are rendered are evocative of later pharaonic representations of the goddess Nut, who was goddess of the night sky and is generally depicted in pharaonic art with an elongated body arched over the Earth. It has been suggested that the identification with Nut may be strengthened by the arrangement of other figures painted in red ochre, adjacent to and beneath the central, white-painted figure. These adjacent figures have been compared to pharaonic representations of Nut's attendant deities.

There are other images at Mestekawi that may have parallels with pharaonic culture and beliefs. Among these is the group shown in plate 20 in which, cleverly, a natural fissure in the sandstone has been used to represent what at first may be interpreted as the reflective surface of a pool of water. Above the fissure are ten figures shown from the waist up as if standing in the pool, with their arms raised above their heads. Inverted images of these individuals are painted below the fissure, but closer scrutiny reveals that these are not reflections; unlike the upright figures with both their arms raised, the inverted figures have their hands and arms in different positions. The position of the arms is not the only difference between the figures above and below the sandstone fissure. Counting from the right, the first five upright figures have an equal number of corresponding, inverted figures. The next group of five upright figures, in the center of the composition, is accompanied by up to nine inverted figures. Farther to the left is a group of about nine inverted figures that have no upright counterpart. Was this composition intended to represent the meeting of this world and the spirit world, with the living represented above the fissure and their spirits shown below? Do the inverted figures that have no upright counterpart represent the spirits of departed loved ones? Bárta certainly considers this to be the case, pointing out that pharaonic Egyptians frequently symbolized the dead by representing them upside down.

Could elements of the Wadi Sura rock art have influenced the development of the pharaonic culture of the Nile Valley? As discussed earlier in this chapter, the latest date established by carbon dating of the Wadi al-Bakht playa deposits (3500 BC) is remarkably close to the accepted start of the pharaonic period in about 3000 BC, particularly when we consider that neither of these dates is precise and may vary by a few hundred years. Unfortunately, however, the most recent occupants of the Wadi Sura area were pastoralists whose distinct rock art does not contain the features that have been identified as possible precursors to elements of pharaonic culture. The use of yellow pigment and the strange headless beasts in Wadi Sura have been dated to an earlier phase of occupation, a period for which we have no evidence for pharaonic or pre-pharaonic activity in the Western Desert.

When standing beneath the looming cliffs of the Gilf al-Kebir or marveling at the beautiful simplicity of the Wadi Sura rock art, it is easy to get caught up in the magic of the Western Desert and fantastic tales of lost cultures that may have inhabited this area. As we have seen when discussing the Bashendi and Sheikh Muftah people (chapter 8) and the people who inhabited the Gilf al-Kebir until the dawn of the pharaonic era, the Western Desert *was* home to distinct cultures that faced environmental pressures associated with natural climate change. As the climate deteriorated and these people migrated out of the advancing desert, those who reached the oases or the Nile Valley would have become subsumed into the existing populations, and it seems inevitable that some elements of their culture would have been adopted by the host population, including cultural beliefs that we now recognize as essentially pharaonic. Is it possible that folk memories of their former lives in Egypt's more remote areas survived in song or other oral traditions? If so, was it these tales of ancestral homelands that led the Early Dynastic Egyptians to explore the far reaches of the Western Desert, culminating in the Old Kingdom with the establishment of regionally important centers such as Balat and the associated network of way stations such as Abu Ballas and Djedefre's Water Mountain to support these desert expeditions? Although we do not currently have a clear understanding of these issues, as further research is undertaken, I feel we will need to re-evaluate our understanding of the indigenous populations of the Western Desert and the role they played in the emergence of the pharaonic culture of the Nile Valley.

The Long Trek Back

The descent from the north of the Gilf al-Kebir is not as dramatic as the ascent from the south. Ground levels fall away gradually as the plateau dips to the north and eventually the rocky surface of the plateau is replaced by the soft sand of the Great Sand Sea. After initially traveling north, parallel with the Libyan border, you turn northeast, aiming for a place called Ain Dalla, an ancient well surrounded by a cluster of palms: life-giving water, and welcome shade from the desert sun. Ain Dalla (fig. 8.1) serves as the gateway to the Farafra Oasis, renowned for its sweeping expanses of pure, white-colored limestones that dazzle the eyes. Resembling either snowdrifts or salt beds, these rocks are 60 to 65 million years old and straddle the boundary between the Mesozoic and Cenozoic eras (fig. 2.1). Their striking white color indicates that at this time, the shallow seas over Egypt were filled with an abundance of marine life, and it is no surprise to find well-preserved fossils of echinoderms (modern sea urchins are echinoderms) and shellfish preserved in these rocks. However, the fascination of the White Desert is not simply due to the color of the rocks or the fossils that they contain.

The White Desert is a bizarrely sculpted landscape. Ground levels would have initially been much higher, with the surface formed by a layer of limestone that was marginally more durable than the layers beneath. Streams and rivers winding over that terrain exploited natural joints in the strata to dissect the landscape into a series of limestone blocks, each one capped by the remains of that more durable surface layer. As the climate became more arid, windblown sand became the principal agent of erosion, imperceptibly removing the softer beds from below and undercutting the remnants of the more durable surface layer. Over the play of geological time, this process sculpted the landscape into a weird forest of mushroom-like pedestals standing upright across the desert floor (plates 5 and 21). In some cases, these top-heavy features have lost their battle with gravity and their upper parts have collapsed (the feature on the ground in plate 21), while in other instances the overhanging upper block rests on its pedestal with just three or four points of contact, seemingly waiting for that final gust of wind to topple it. A few hours spent in the White Desert is enough to underline the intense aridity of this area. The heat, the fine white dust, and the sand soon begin to make their presence felt. The White Desert is not the place to be in a sandstorm.

The Faiyum Basin

As you might recall from chapter 1, the Faiyum basin is a large bowl-shaped depression that in the recent geological past has been occupied by a large freshwater lake, known today as Birkat Qarun. The strata exposed in Faiyum are Eocene or younger (fig. 2.1) and their significance has been recognized by paleontologists since the 1880s, given that they contain some of the richest fossil deposits in Egypt. In addition to petrified wood (fig. 5.1) and fossilized whales and turtles similar to those at Wadi al-Hitan (plate 7), there are a range of other fossils that have been found, including 9-meter snakes, birds, and giant mammals such as *Arsinoitherium* and *Hyaenodon*. Perhaps the most important fossil group from Faiyum, however, is the primates, with thirteen different species having been identified in the area, many of which are unknown anywhere else in Africa.

The hydrogeology of the Faiyum basin is different from that of the other depressions in the Western Desert. Unlike the oases, the base of the Faiyum depression does not reach the Nubian Sandstone aquifer, and so, unlike the oases, the fertility of Faiyum relies on surface water from the Nile, which enters via the Bahr Yusuf. Although Bahr Yusuf translates as "Joseph's Canal," this is clearly a natural watercourse, an abandoned section of the Nile that originally left the river near Asyut, to run north for more than 250 kilometers before passing through the Hawara Channel and entering the Faiyum depression. Evidence from a series of raised beaches surrounding Birkat Qarun indicates that initial connections with the Nile were established during a prolonged period of high Nile floods, which most likely occurred at some point late in the Prenile phase of the river (chapter 6). These initial connections appear to have been relatively brief and were followed by a period during which there were no links with the river and lake levels fell. It was not until as recently as perhaps 9,000 or 10,000 years ago that a

Figure 9.10. The towering ruins of the mudbrick walls that once surrounded the temple area at Dimeh al-Siba.

subsequent series of sustained high Nile floods is thought to have broken through the Hawara Channel to once more flood the depression. These more recent connections with the river are thought to have been sustained until about 8,000 years ago and were followed by a complex series of fluctuations in lake levels, nearly leading to the complete desiccation of the basin. Connections with the Nile appear to have been re-established from about 7,000 years ago, and it is from this time that we have evidence for human activity in the Faiyum region, particularly along raised beaches to the north of the lake, where early settlers made temporary or seasonal camps, utilizing the freshwater for fishing. These settlements also provide evidence that the communities there had adopted early methods of agriculture.

The Faiyum lake was known as Lake Moeris by the people of pharaonic Egypt, and the water levels continued to fluctuate heavily throughout their time. The lake was probably at its greatest during the Old Kingdom (fig. 1.6), with water filling most of the Faiyum depression; however, a period of repeated low Nile floods led to the drying up of the Hawara Channel and it became necessary for the pharaohs of the Middle Kingdom to clear the channel to re-establish flow from the river. Later, during the Greco-Roman period, engineering works allowed the flow of water through the Hawara Channel to be controlled, enabling the lake level to be lowered to expose additional areas for agriculture. These measures to control water levels and reclaim land for agriculture have served the area well, with the Faiyum remaining one of the most agriculturally productive areas in Egypt today. The lowering of levels in Lake Moeris, however, had a number of significant consequences, one of the most dramatic examples being revealed by the remains of the Greco-Roman port city of Dimeh al-Siba, which now lies more than 2 kilometers from the current lake shore. Only the ruins of the innermost part of the city can be seen today, largely built from mudbrick but with a single stone-built temple. By far the most impressive remains at Dimeh are the precarious ruins of towering mudbrick walls that surrounded the city's inner temple precinct and even now, in their ruinous state, reach perhaps 10 meters high (fig. 9.10).

Asphalt roads quickly lead northeast, away from the Faiyum toward Cairo, taking you out of the otherworldly landscape of the Gilf al-Kebir, the Great Sand Sea, and the White Desert to the modern, vibrant, and chaotic hubbub of Egypt's capital. Although exploration of the desert regions is difficult, it is clear from a raft of recent and fascinating discoveries that such exploration is vital if we are to develop our understanding of the shifting landscape west of the Nile Valley, and the people who lived there many thousands of years ago.

CHAPTER 10
MINING AND QUARRYING

A great deal of research has gone into understanding the history of mining and quarrying in ancient Egypt. One of the most remarkable features established by this research is the extent to which stone (particularly decorative stones), metals, gemstones, and other mineral deposits were exploited throughout the pharaonic era. As we will see in this chapter, not only did the ancient prospectors have a surprisingly broad understanding of the distribution of mineral wealth across the landscape, the lengths to which teams of miners and quarry workers went to exploit those mineral deposits is truly astonishing.

Exploiting the Mineral Wealth of the Eastern Desert

Although quarrying and mining were undertaken in many parts of Egypt during the pharaonic era, by far the most important area in terms of the variety of metals, gemstones, and other available natural resources was the Eastern Desert (fig. 10.1). The source of most of this rich mineral wealth was the Precambrian strata of the Red Sea Hills. As we saw in chapter 3, over the many millions of years since their formation, the Precambrian Basement strata have been

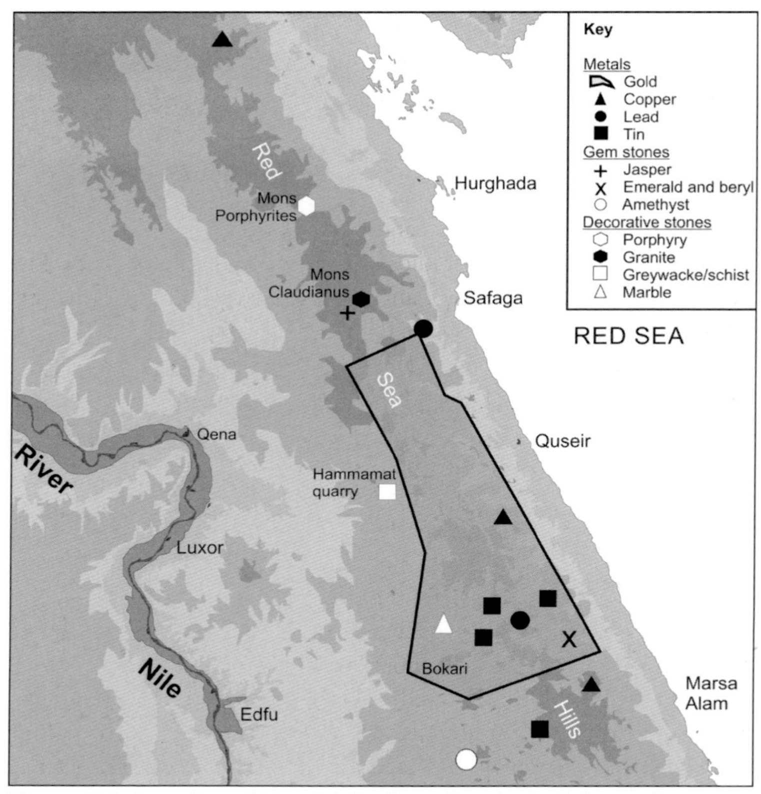

Figure 10.1. Overview of the distribution of ancient mining and quarrying in the central part of the Eastern Desert.

subject to a wide variety of tectonic processes as they have been dragged across the surface of the Earth. These processes have led to the development of a complex range of igneous and metamorphic rocks within the Basement strata, together with the formation of mineral ores and gemstones. Across most of Egypt, this mineral wealth is largely inaccessible, being buried beneath substantial thicknesses of Mesozoic sandstones and Cenozoic limestones. In the Eastern Desert, however, the opening of the Red Sea Rift in the geologically recent Oligocene Epoch (chapter 5) led to sections of Precambrian Basement being thrust up to form the Red Sea Hills.

After weathering and erosion removed the overlying limestones and sandstones from these mountains, the ancient mineral-rich Precambrian strata were exposed.

Metals

Establishing even an approximate date for the start of mining and quarrying in a particular area can be difficult because later workings can often obscure or even eradicate evidence for earlier activity. The presence of *serekh*s of some of ancient Egypt's earliest pharaohs (fig. 7.8 left), however, suggests that the people of the Nile Valley were actively seeking to exploit the mineral wealth of the Eastern Desert from a very early stage. The earliest pharaonic mining and quarrying expeditions may have relied on the indigenous people of the area, both to locate areas of workable deposits and as a source of labor during extraction and processing. Copper is generally recognized as the first metal to be used by early man and it appears to have been discovered independently by a number of early civilizations. Copper often occurs in its native form, that is, in the form of metal nuggets that early cultures may simply have encountered lying on the ground. Early prospectors seeking copper may also have identified visual clues in the landscape associated with the distinctly colored weathering products of copper minerals such as malachite and azurite.

There was a significant demand for copper in the early pharaonic era, particularly in the Old Kingdom. It is remarkable to consider that tools made from this relatively soft metal quarried and shaped everything from the blocks of limestone used to build the pyramids to the statues that decorated ancient temples. Detailed analysis by Denys Stocks (see Further Reading) has shown that by alloying copper with small proportions of other metals and possibly by applying a range of processes such as tempering during casting or hammering, the durability of ancient copper tools could be enhanced. Stocks has also demonstrated that by combining the use of tubular copper drills with sand or other quartz grit as a cutting agent, it was possible to work a range of durable stones including granite.

Given that copper is often found occurring naturally with other minerals, it can be difficult to determine whether the presence of other metals in ancient copper tools was deliberate or accidental.

Evidence suggests that in the Old Kingdom and possibly earlier, arsenic was being added to copper to produce a form of bronze. The first intentional use of copper/tin alloys in Egypt is not thought to have occurred until the Middle Kingdom (fig. 1.6); however, evidence from hieroglyphic graffiti in the Eastern Desert suggests that the pharaonic Egyptians may have been seeking out tin ore from as early as the Old Kingdom. As discussed in chapter 7, graffiti from the pharaonic era are an important aspect of the rock art of the Eastern Desert, particularly in the ancient quarrying areas. These texts often include prayers for the success of an expedition, or the names of the expedition leaders themselves. Many texts make it clear that gold was their main objective. In one particular wadi, however, Old Kingdom graffiti has been identified extending further east beyond the known gold-mining areas, deeper into the wadi system. Where the Old Kingdom graffiti ends, Russell Rothe and his colleagues have found specimens of the black tin ore cassiterite among the wadi gravels (see Further Reading). If these Old Kingdom expeditions were not seeking tin ore, it is difficult to identify any other reason for them to have traveled to such remote parts of the wadi system. Clearly, however, even if the Old Kingdom prospectors had been exploiting tin, this does not indicate that tin was being alloyed with copper at that time.

As we saw with native copper, the earliest exploitation of gold may have focused on the collection of what are referred to as placer deposits, nuggets weathered out of hillsides or washed along watercourses. Because gold is very stable and tends not to combine with other elements, early prospectors will not have been able to identify the presence of gold by means of brightly colored weathering products, as was the case with copper, for example. Other than placer deposits, therefore, there would have been few obvious signs of the presence of "the flesh of the gods" in the Eastern Desert. Just like copper, however, much of the gold mineralization of the Red Sea Hills would have been associated with quartz veins (fig. 3.2), and once a link between quartz and brightly colored copper ore had been made, it is likely that the presence of gold would have also been noted as the veins were worked.

As figure 10.1 indicates, there was widespread gold-mining activity in the Eastern Desert, some of which began as early as the

Predynastic Period. The site of Bokari (fig. 10.1) was in use from at least the Old Kingdom right through to the Greco-Roman period and in many ways represents a typical Eastern Desert gold-mining site. The focus of activity at Bokari was a cluster of buildings roughly constructed from locally available blocks of stone (fieldstones) stacked together without the use of mortar (fig. 10.2). These would have included buildings for administration, accommodation, kitchens, and workshops. The mineral workings themselves extended some considerable distance away from the central encampment and focused on the extraction of outcropping quartz veins (fig. 10.3), which would be worked to quite substantial depths until lack of oxygen in the excavation made further mining impractical. If necessary, parent rock would be removed from alongside the vein so that the quartz could be extracted to greater depth. Although relatively rare, underground mining was used at a small number of sites in the Eastern Desert, to fully extract a particularly rich gold-bearing vein. Once broken out of the vein, the barren quartz was left at the side of the excavation (fig. 10.3) and the gold-bearing material taken back to the main camp for processing.

Until the end of the Middle Kingdom, gold processing appears to have been limited to the use of hammer stones to crush the quartz, with the flakes of gold picked out by hand. Some early gold-mining operations may have been undertaken by indigenous people, perhaps as part of trading arrangements made with the

Figure 10.2. The main encampment at Bokari.

Figure 10.3. A quarried quartz vein at Bokari. The "barren" quartz, which the experienced quarry workers will have ensured contained no minerals, has been left at the lower end of the trench (arrowed).

people of the Nile Valley. The New Kingdom, however, saw something of an "industrial revolution" in the Eastern Desert, implying that from this time, mining and quarrying operations were fully in the hands of the pharaonic Egyptians. The earlier use of hammer stones was replaced by more efficient milling techniques that had originally been developed to grind grain, and although there are indications that washing processes may have been used at earlier times, washing became more widely adopted in the New Kingdom. It is thought that the origins of the legendary Golden Fleece may lie in gold-washing operations such as these. Woolen fleeces laid on the mortar-covered surface of specially prepared stone-built washing tables would have trapped gold particles as the less dense quartz fragments washed over. Once the fleece had collected sufficient gold, it was probably burned to isolate the precious metal.

The New Kingdom also saw a significant increase in the scale of gold prospecting across the Eastern Desert, with the prospectors appearing to have developed a much greater understanding of the way in which the geology of the region influenced the natural distribution of gold. This included not only identifying new sites for mining gold but also an understanding of the processes of weathering and erosion, with evidence that wadi gravels were being systematically worked for placer deposits. Remarkably, some of the work of these New Kingdom prospectors has been preserved in the form of a 3,500-year-old papyrus map that is now held in the Egyptian Museum in Turin and has come to be regarded as the world's oldest surviving geological map. The map depicts an area of Wadi Hammamat (fig. 7.2) and in addition to documenting important areas of gold mining, also identifies areas of ancient *bekhen*-stone quarrying (see below). Much like modern geological maps (plate 3), the ancient map uses colors and a key to identify the different soil and rock types in the area, together with notes added in hieratic, a simplified form of hieroglyphs. Remarkably, features identified on the map can be recognized on the ground today.

Decorative stones: The *bekhen*-stone quarries of Wadi Hammamat

Researchers struggle to agree on the correct geological term for the *bekhen*-stone exposures of Wadi Hammamat, with the stone described variously as schist or greywacke. Strictly speaking, the actual range of stones quarried from Wadi Hammamat included siltstone, greywacke, and conglomerate. *Bekhen* and similar rock types were used throughout the pharaonic era for quite specific uses. In the Early Dynastic Period, *bekhen* was used to produce a range of objects including palettes, thought to have been used for the grinding of cosmetics. Many of the palettes that have survived, including the Narmer Palette, named after one of the first kings of the First Dynasty (fig. 11.3), are thought to have been used for ceremonial purposes and are generally considered to be too large to have been used in any practical manner. In the Old Kingdom, *bekhen*-stone was also used for statues, including the wonderful Menkaure triads, a series of statues that were recovered largely intact from this pharaoh's pyramid complex at Giza. Each of these statues is a masterpiece, showing the king accompanied by a queen and a female deity (fig. 10.4). In later periods, Egypt's elite favored burial in exceptionally crafted sarcophagi, some of which were carved with the greatest skill from *bekhen*-stone. We may never fully understand what it was about *bekhen*-stone that drove the ancient Egyptians to quarry from such a remote location as Wadi Hammamat, but key factors are likely to include the ability of the stone to retain a great deal of fine detail, the deep, lustrous polish that could be achieved, and the color. The weathered surfaces of the *bekhen*-stone quarry have developed a rather unremarkable dark gray patina, but beneath this patina, much of the rock displays a rich, honey-gold tone.

Within the ancient *bekhen*-stone quarries of Wadi Hammamat, only limited quarrying appears to have taken place along the southern side of the wadi. As a result, hundreds of hieroglyphic and hieratic inscriptions have survived on the exposed rock faces. While some of these inscriptions are little more than graffiti, many of the illustrations and texts would grace any museum collection. A number of the surviving inscriptions are complete, recording the feats that the ancient quarrymen accomplished on behalf of their pharaoh and the gods, often describing the great personal honor that was felt by the leaders of these quarrying operations. The working areas of the Hammamat quarry appear to have been extended from the northern side of the wadi, and among the quarry rubble and scree are the remains of stone-built huts used by the quarrymen and a number of unfinished sarcophagi. In one area near the western end of the quarry, high on the northern wall, is by far the most accomplished piece of rock art in the entire quarry area. Although damaged, this rendering of the New Kingdom pharaoh Seti I offering lotus flowers to Amun-Re (plate 22) has been carved in low relief. As so often with art from this period of the New Kingdom, the quality of this carving is superb, with particular features such as Amun-Re's ear sculpted in incredible detail. Also noteworthy is the clever way the artist has worked with the patination of the stone to

Figure 10.4. One of the magnificent *bekhen*-stone statues recovered from the Old Kingdom pyramid complex of the pharaoh Menkaure at Giza. Although the quality of these carvings owes much to the skill of the craftsman who worked on it, credit must also be given to the selection of stone from which these statues were carved. The fine-grained nature of the *bekhen* stone allowed incredible detail to be added to the carvings.

reproduce the overlapping lower section of the god's beard and the decorative collars that adorn both figures. The overall composition of this remarkable piece is undoubtedly enhanced by the natural golden color of the stone.

Decorative Stones: The granite quarries of Mons Claudianus

Although the scale of the pharaonic quarrying operations at Wadi Hammamat is impressive, arguably these achievements were surpassed by the Imperial Roman granite (more strictly granodiorite) quarrying operations at Mons Claudianus. Perhaps most startling is the immense logistical effort that was required to support these operations in this extremely remote part of the Red Sea Hills (fig. 10.1).

Although there is little in figure 10.5 to give an indication of scale, by comparison with most settlements in the Eastern Desert, the main camp at Mons Claudianus is huge, with a walled area that accommodated over a thousand people. Within the walled enclosure, it is possible to walk the ancient streets and even to enter some of the individual buildings with their stone roofs still intact. Next to the main camp is a second sizable building which was constructed as stables for the draft animals that were used not only as part of the quarry operations, but also for hauling the multi-wheeled carts required to transport the quarried stones across the vast wilderness of the Eastern Desert to the Nile Valley and eventual shipment overseas. The columns that grace the entrance to the Pantheon in Rome, for example, were quarried at Mons Claudianus.

The quarry was in use from the late first to the mid-third centuries AD and the active quarry area extended for many miles around the main camp. In addition to working areas in which blocks of granite can still be seen in various states of preparation before being detached from the bedrock, everywhere lie the abandoned remains of columns, column bases, and column capitals, together with stone-built ramps that were used to load carts for onward shipment of the quarried goods. Large areas of the extended quarryscape have the appearance of stockyards, in which semi-prepared items sit together, grouped by type. The most impressive object at Mons Claudianus is an abandoned column of truly enormous proportions (fig. 10.6). It is clear that during the quarrying of this column, flaws

Figure 10.5. The main fortification at Mons Claudianus.

Figure 10.6. Deep in shadow from the strong desert sun, a gigantic column lies abandoned at Mons Claudianus. With an estimated weight of 200 tons, this column was abandoned because fractures were revealed in the rock as it was being quarried.

were noticed in the rock, and it appears that clamps were applied in an attempt to prevent these flaws developing into something more significant. These measures were not sufficient, however, and the column split at a number of places before eventually being abandoned. On the basis of other column fragments nearby, it seems that it was relatively common for large items such as this to fail during quarrying, and rather than waste the considerable effort already expended, damaged items were often recut into smaller blocks.

Gemstones and other natural resources

The geological processes that led to the formation of the many varieties of precious and semiprecious stones (gemstones) are discussed in chapter 3. As indicated in fig. 10.1, there is clear evidence in the Eastern Desert for ancient mining and quarrying of gemstones such as jasper, beryl, and amethyst. Given that gemstones typically occur in small crystal masses and thin veins, many of the known quarry sites are relatively small and it is likely that many others remain so far unidentified. As was the case with quarrying for metal ores, the methods used to extract gemstones were generally based on the use of shallow pits and trenches; however, in some cases such as emerald mines (a form of beryl), underground workings were used. Extensive copper workings have been found in Sinai (fig. 1.1), together with ancient turquoise mines. Turquoise is a mineral that contains copper and aluminum and was prized by the pharaonic Egyptians for use in jewelry as well as the highly decorative metalwork that has become an iconic feature of pharaonic art (fig. 1.7). Although the earliest identified sites in the Sinai are from the First Dynasty, it is likely that there was a much older tradition of mining and quarrying in the area.

Ancient quarrying activity in the Eastern Desert was not restricted to metals, decorative stones, and gemstones. In the foothills of the Red Sea Hills, a little distance east of the Jacuzzi (fig. 7.1), are the remains of a small cluster of stone buildings inside of which are piles of green-gray, platy stone fragments. This unremarkable-looking material is natural talc. Talc veins are common in metamorphic strata, and a quick survey of the slopes surrounding this small settlement identified a number of partially worked-out talc seams, indicating that this small and rather remote site had been established specifically for the extraction of this single natural resource. Judging by the ceramic fragments that were scattered across the site, these talc workings had been active in the Greco-Roman period, illustrating that in addition to metals and gemstones, the ancient Egyptians were willing to endure the hardships of the Eastern Desert for more everyday natural products.

Building Stone

Many structures in ancient Egypt were built using a core of locally quarried stone (building stone), which was then cased with an outer layer of more decorative stone that typically needed to be imported to the site. Perhaps the best-known example of this approach is at the Old Kingdom necropolis in Giza, where a series of massive quarries were excavated in the outcropping Eocene limestone plateau for masonry used in the core of the pyramids and associated temples and tombs (fig. 10.7). For the sake of efficiency, the quarries at Giza were located as close as possible to the pyramids, and although they have become largely infilled with sand, sections of ancient quarry walls can still be seen in some of the less visited parts of the Giza necropolis, characterized by their distinctive pattern of diagonal markings, formed by the action of the tools used to quarry the stone (fig. 10.8). Despite these quarries being a vital component of the development of the Giza Plateau, their significance is generally overlooked. For example, it has been frequently observed that the three main pyramids at Giza were built along a diagonal line running from northeast to southwest, with some researchers claiming that this alignment was intended to represent certain stars in the night sky. Sadly, the truth may be far more mundane. It appears that the diagonal arrangement of the Giza pyramids may simply reflect the topography of the site together with the evolving quarryscape, with the site of each successive pyramid influenced by the location of earlier quarries as well as a large wadi that defines the southern limits of the Giza Plateau.

Except perhaps for the earliest step pyramids at Saqqara (see chapter 11), the limestone used to case ancient Egyptian pyramids

Figure 10.7. The Great Sphinx and the pyramid of Khafre, with the remains of ancient quarries behind. These large quarries (arrowed) were for the limestone masonry used to build the core of the pyramids and are largely infilled with sand.

Figure 10.8. The walls of a limestone quarry at Giza, showing clear diagonal quarry marks. The stone extracted here was used to build the pyramids, temples, and other tombs of the Giza necropolis.

was quarried from both opencast workings and enclosed galleries at Tura at the southern tip of the Nile Delta, near ancient Memphis (fig. 11.1). This fine white limestone was quarried throughout the pharaonic era and, remarkably, quarrying is still underway at Tura today. Like the Tura quarries, there has been a long tradition of quarrying at Gebel al-Silsila in southern Egypt (fig. 11.1). In this case, however, sandstone was extracted for masonry for temples such as those at Luxor and Karnak (ancient Thebes, fig. 11.1). Given the size and extent of these ancient building projects, it should be no surprise that the ancient quarry workings at Gebel al-Silsila are enormous.

In addition to limestone and sandstone for masonry, a variety of other stone types were used in construction throughout the pharaonic era, including basalt, travertine, quartzite (a form of metamorphosed sandstone often used for large statues), and granite. With the exception of basalt, which appears to have been quarried at Widan al-Faras, north of the Faiyum region (fig. 11.1), the main quarry sites for these stones were located close to the Nile, which aided the movement of the quarry product by river to the construction sites. Many of these stones had specific uses. Basalt was often used to pave causeways and pyramid mortuary temples (fig. 10.9), most likely because its black color was associated with *kemet*, the Black Land (see chapter 11). Travertine (often referred to as Egyptian alabaster), generally quarried at Hatnub (fig. 11.1), was often used to pave valley temples, while the granites at Aswan were quarried for use as casing stone for buildings as well as for statues, columns, and the mighty obelisks.

The Unfinished Obelisk

By far the largest obelisk ever attempted by the ancient Egyptians remains unfinished in its quarry in Aswan. If completed, the Unfinished Obelisk (fig. 10.10) would have stood some 42 meters high and would have weighed about 1,200 tons. It may have been commissioned in the New Kingdom (fig. 1.6); however, some researchers have suggested a Late Period date for it. The Unfinished Obelisk lies several meters below the surface of what may have originally been a rounded granite hill, and several meters of granite overburden needed to be removed before quarrying of the obelisk could begin. Although quarrying at this location may have been commissioned specifically for the Unfinished Obelisk, it is clear

Figure 10.9. Black basalt used to floor the pyramid temple of the Great Pyramid at Giza. The lower parts of the Great Pyramid can be seen in the background. The use of this dark-colored stone is thought to have symbolized *kemet*, the Black Land.

Figure 10.10. The Unfinished Obelisk at Aswan. Note what appear to be cracks in the granite running through the upper section of the monolith.

that quarrying continued, with the scars of a number of smaller obelisks that were completed and removed from the quarry still evident. The Unfinished Obelisk does not lie horizontally in the quarry, with the base sitting about 6 or 7 meters lower than the apex. Evidence from a series of ancient, hand-dug, vertical shafts that surround the Unfinished Obelisk and were excavated from the original summit of the hill suggest that the angle at which the obelisk lies in the quarry may have been carefully selected to take advantage of the best-quality granite at this location. Modern construction projects tend to begin with a program of investigation, which often uses boreholes to understand the ground conditions. Remarkably, the ancient shafts that surround the Unfinished Obelisk make it clear that before quarrying work began, the individuals charged with producing what would have been the largest obelisk in ancient Egypt took similar steps to assess whether ground conditions at this location appeared favorable.

Once the location was fixed and the unsuitable overburden had been removed, a deep trench was cut to form the outline of the obelisk. From the evidence littering the site, the majority of this quarrying and trenching work was undertaken using dolerite hammer stones (fig. 10.11), naturally rounded pieces of a particularly hard igneous rock, which would have been held in two hands and repeatedly struck against the exposed granite. This hammering process has left characteristic scars both on parts of the obelisk and along the floor of the surrounding trench (fig. 10.11). There is some evidence that fire may have been used to aid granite quarrying; however, it is not clear how widespread this practice was. By burning the exposed granite surface and then dousing it with water, the upper surface of the granite would become more brittle, allowing for an increased rate of extraction using hammer stones. Once the trench surrounding an obelisk reached the required depth, the bedrock was quarried away from beneath the obelisk until all that connected it to the bedrock was a narrow granite rib. Presumably, once the rib was thin enough, levers would be used to break the obelisk free from the quarry. In the case of the Unfinished Obelisk, however, work never reached that final stage.

Despite the evidence that a great deal of careful preparation was undertaken before quarrying for the Unfinished Obelisk began, some researchers suggest that the project was abandoned because the granite started showing evidence of cracks, fatal flaws that meant as soon as an attempt was made to lift the obelisk, it would break. As the accompanying photographs show, there are several fissures that cut across the structure; however, in most cases close inspection suggests these may not be natural. The wide and rounded form of these fissures is not consistent with what would be expected for natural discontinuities and they do not appear to penetrate to a significant depth into the rock mass. Given the shallow, rounded nature of these fissures, it seems more likely that they

Figure 10.11. Characteristic "pockmarked" surfaces on the tip of the Unfinished Obelisk and in the trench surrounding it. These marks were probably made by the use of dolerite hammer stones for most of the quarrying activity. Note a hammer stone lying in the trench to the right of the obelisk.

are part of later attempts to break up the Unfinished Obelisk to produce smaller items from the abandoned granite block. If my interpretation is correct, it would suggest that the Unfinished Obelisk was abandoned not because of fractures in the stone, but because on this occasion, the ancient Egyptians had simply overreached themselves. As the obelisk neared completion, it perhaps became apparent that the logistical challenges that would need to be overcome to move this huge piece of granite were too great. Manpower may have been relatively unlimited, but in the confines of a newly opened quarry, there may have been insufficient space to use the available manpower effectively. What was intended as the biggest obelisk Egypt had ever seen may simply have turned out to be too ambitious.

The natural tendency for anyone with an interest in ancient Egypt is to focus on the incredible buildings that line the Nile Valley, from the Temple of Isis, at Philae just south of Aswan, to the incredible pyramids near modern Cairo. Alternatively, enthusiasts and non-enthusiasts alike might marvel at the remarkable craftsmanship from which some of the world's most iconic pieces of art were produced, such as the funerary mask of Tutankhamun (fig. 1.7). What I hope to have demonstrated with this brief summary of ancient mining and quarrying operations are the great lengths to which the people of ancient Egypt needed to go in order to provide the raw materials to support the builders, metalsmiths, and craftsmen of the Nile Valley. If it had not been for the vast natural wealth of the Egyptian landscape, or the ability to send organized expeditions out into the wilderness of the Red Land to exploit the resources of remote areas such as the Red Sea Hills, it is certain that the civilization of the River Nile would have been very different from the civilization of ancient Egypt that is celebrated so widely today.

Plate 1. A golden sunset at the Giza pyramids.

Plate 2. The dramatic cliffs at Deir al-Bahari, with the mortuary temple of Hatshepsut nestling to the lower right. The triangular peak to the upper left—al-Qurn—overlooks the Valley of the Kings behind (see chapter 11).

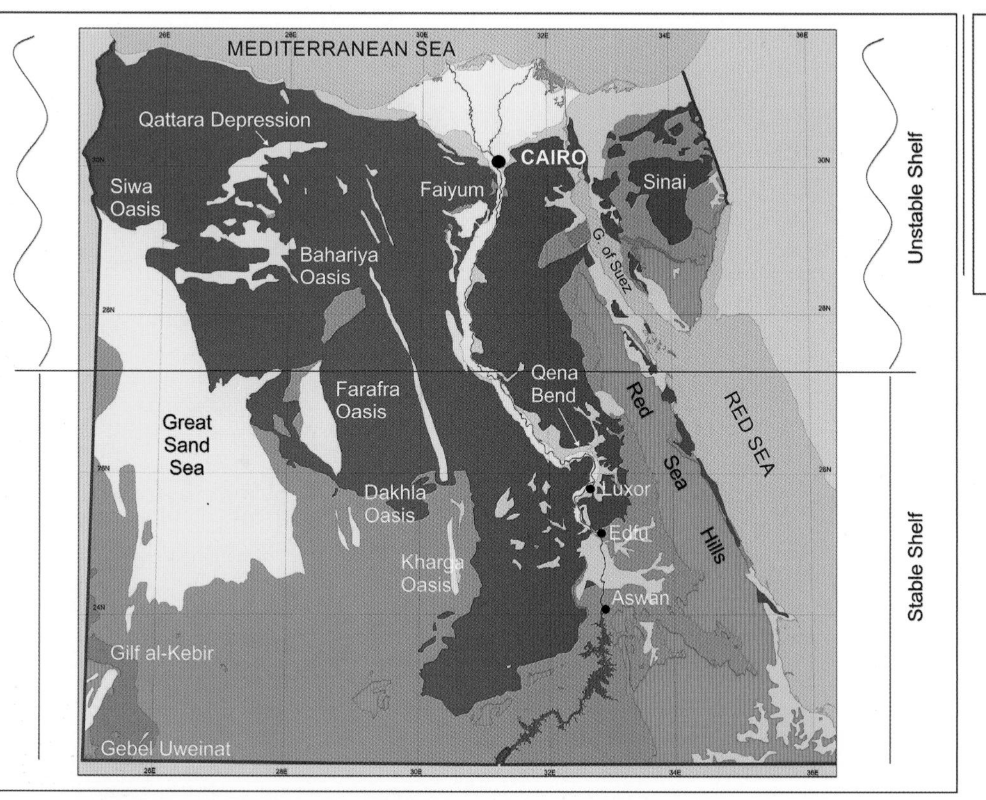

Plate 3. A simplified geological map of Egypt.

Plate 4. Precambrian conglomerates in the Eastern Desert.

Plate 5. The bizarre and spectacular landscape of the White Desert, Farafra. These calcium carbonate-rich rocks straddle the boundary between the uppermost Mesozoic strata and the lowest Cenozoic rocks of Egypt and were deposited in shallow, warm-water conditions some 55 to 65 million years ago.

Plate 6. The Great Sphinx of Giza with the Great Pyramid of Khufu (right) and the pyramid of Khafre (left). The alternating sequence of thin sandstones and siltstones that can be seen across the chest of the Sphinx reflect conditions in a near-shore lagoon that existed in the Eocene, some 40 million years ago.

Plate 7. Wadi al-Hitan. The fossilized skeleton of *Dorudon atrox*, an early whale that inhabited the ancient seas of northern Egypt, preserved largely in the position in which it died some 40 million years ago.

Plate 8. Gebel al-Marsous, a basalt-capped hill in the Western Desert, showing detail of the columnar structure formed in the basalt as it cooled from a molten state (inset).

Plate 9. Although covered in dust from their dry desert surroundings, the large crystalline formations at Crystal Mountain are spectacular.

Plate 10. Poised at the crest of a relatively small dune in the Great Sand Sea. Twenty million years ago this area is likely to have been occupied by densely forested river valleys.

Plate 11. A geological unconformity in the Eastern Desert to the west of the Red Sea Hills. In the left of the picture, the dark Precambrian Basement rocks still retain a remnant cover of much younger Mesozoic Nubian Sandstones. To the right, the overlying sandstones have been completely weathered away.

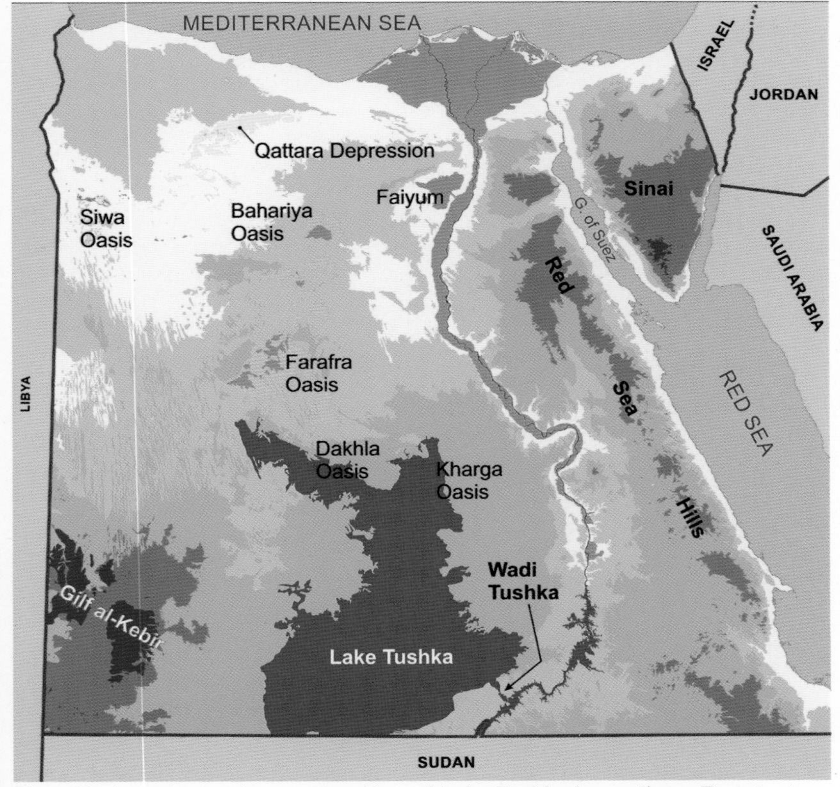

Plate 12. A conjectural reconstruction of Lake Tushka in southern Egypt, at possibly its greatest extent during the Protonile phase of Nile evolution.

Plate 13. The Great Pyramid of Khufu on the Giza Plateau, just outside modern Cairo. Built about 2500 BC, on the top of the Eonile cliff and therefore out of the reach of the Nile's annual flood, this is the most perfectly executed of over 100 pyramids that were built in Egypt during the pharaonic period. Until recently, the odd-shaped white building at the foot of the pyramid (to the right) housed an ancient wooden boat from the time of Khufu, which was buried in a pit next to the pyramid. The lower white building to the left sat over a second ancient boat burial.

Plate 14.. A sandy wadi in the Eastern Desert. These wadis are former river channels that, for tens of thousands of years before the onset of today's climate, drained vast areas to the east of the Nile.

Plate 15. A mature tree in an Eastern Desert wadi.

Plate 16. The western foothills of the Red Sea Hills.

Plate 17. Libyan Desert Glass (LDG), a form of natural silica glass with an unusually high silica content. Fragments of LDG can be found in a relatively small area of the Great Sand Sea.

Plate 18. A pectoral from Tutankhamun's burial. The scarab that forms the centrepiece of this important piece of funerary regalia is carved from Libyan Desert Glass.

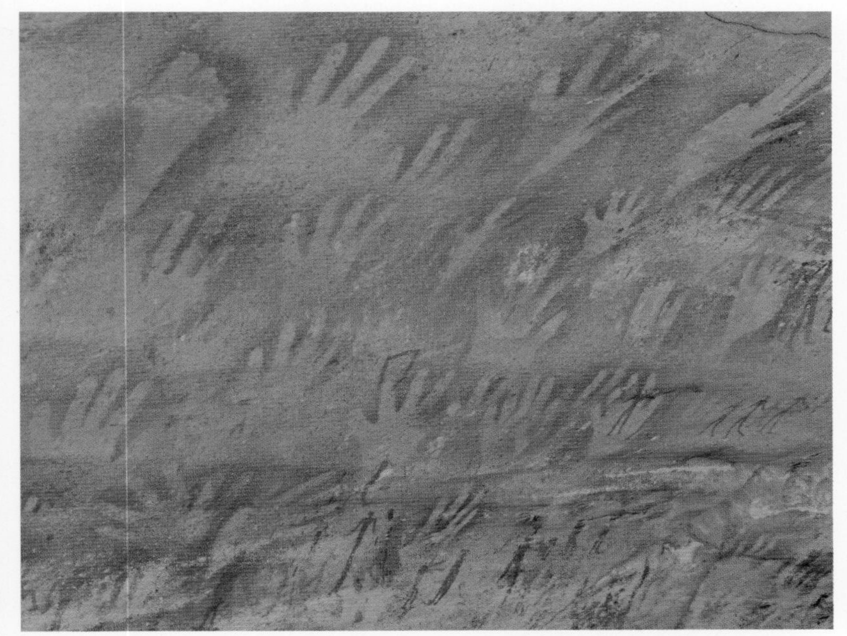

Plate 19. Mestekawi Cave. Countless painted silhouettes of hands line the rock shelter at Mesekawi. Most appear to be waving but one individual (top left) chose to shake their fist at posterity.

Plate 20. Mestekawi Cave. Early mysticism? At first this group, particularly the five figures to the right, appear to have been painted with reflections—perhaps with the fissure in the sandstone used to represent the surface of a pool of water. Look again however and what is painted below the fissure is clearly not a reflection.

Plate 22. The New Kingdom pharaoh Seti I offering lotus flowers to Amun-Re. Low-relief carved decoration on the north wall of the *bekhen* quarries in Wadi Hammamat.

Plate 21. The White Desert. The rocky pedestal at the left has collapsed and now lies on the desert floor.

Plate 23. The early Third Dynasty Step Pyramid of Djoser at Saqqara.

Plate 24. The three great pyramids of Giza. From left to right, Menkaure, Khafre, and Khufu. Due to perspective and the vast size of the Giza necropolis, the Menkaure pyramid looks much larger, relative to the pyramids of Khafre and Khufu, than is actually the case.

Plate 25.. An unfinished granite lintel from Tanis. Detailed study of incomplete hieroglyphic inscriptions like this may tell us a great deal about the stoneworking techniques used by ancient craftsmen.

Plate 26. The sparse vegetation at the northern end of the Wadi Abusir. The Fifth Dynasty pyramids of Abusir lie in the background.

CHAPTER 11
BUILDING IN STONE IN ANCIENT EGYPT

Having looked at mining and quarrying in the last chapter, it seems appropriate to address what we know about the use of stone in construction in ancient Egypt. After all, the awe-inspiring pyramids and the great temples of the pharaonic era are now as much a part of the landscape of Egypt as the Eonile Canyon, the Red Sea Hills, or the Great Sand Sea.

There are many publications that address construction in ancient Egypt and do so in a far more comprehensive way than I can achieve in just a single chapter (for example, Dieter Arnold's *Building in Egypt*; see Further Reading). I have therefore focused on a number of key issues, including the early development of pharaonic construction in stone, pyramid building, and the construction of tombs in the Valley of the Kings. To close this chapter, I have also looked at one of pharaonic Egypt's most enduring mysteries: the precision that was achieved when carving hieroglyphs into hard stones such as granite.

The Earliest Uses of Stone

The pharaonic Egyptians were not the earliest culture to use stone in construction, as is often thought. Perhaps the earliest recognized

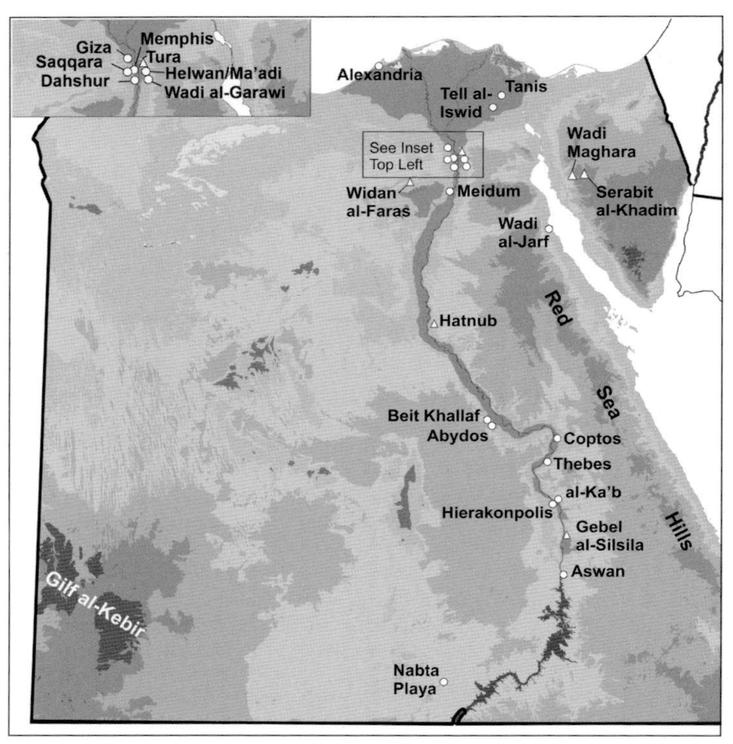

Figure 11.1. Map of Egypt showing some of the key quarrying and archaeological sites referred to. Triangles are quarry sites.

use of stone dates from approximately 11,000 years ago at Göbekli Tepe in Turkey, where a series of circular structures have been found that contain T-shaped pillars up to 5 meters tall. Unlike many early stone masonry elements, these pillars had been carefully and deliberately shaped, and many were decorated in stylized animal reliefs. Slightly younger, dating to about 10,000 years ago, is the Tower of Jericho, an 8.5-meter-high masonry structure that contained an internal stone staircase. The earliest stone-built structures in Europe date from about 6,000 years ago (c. 4000 BC) and include the burial structures at West Kennet in England and Newgrange, Ireland, along with a series of temples on the islands of Malta and Gozo. Henges such those at Avebury and Stonehenge in England were built after 3000 BC.

Figure 11.2. The stone circle discovered at Nabta Playa. It has been suggested that this is a calendar circle, with possible alignments to the summer solstice and significant stars.

Probably the earliest stone structures known in Egypt are from Nabta Playa (fig. 11.2). As discussed in a paper by Fred Wendorf and Romuald Schild (see the conference proceedings edited by Renée Friedman in Further Reading), this low-lying area west of the Nile Valley was occupied by a lake during the Holocene Wet Phase (see chapter 6) and served as a focus of activity for nomadic and semi-nomadic cultures that appear to have had sub-Saharan origins. The area surrounding the lake was occupied on a largely seasonal basis, and by about 8,000 years ago these cultures appear to have developed a religious or cult system that focused on cattle. Sacrificed cattle have been found at a number of locations, buried in clay-lined chambers capped with stone. At one site, upright stones were erected in roughly circular arrangements around a series of pits in which other, larger stones had been buried. The buried stones weighed up to two and a half tons each and were roughly fashioned to resemble cattle. It has been suggested that this cattle cult is perhaps a forerunner of the pharaonic cult of Hathor; however, this link has not been established with any certainty. Sometime later (about 7,000 years ago) a stone circle was erected at Nabta Playa, which some consider to be a calendar circle, with elements aligning with the summer solstice and a number of significant stars. Because the remote location of Nabta Playa makes it difficult to protect the site, many of the excavated stones have been transferred from their original position (fig. 11.2) to the Nubian Museum in Aswan.

Precursors to the Use of Stone in the Nile Valley

Except perhaps for Nabta Playa, we currently have little clarity when it comes to understanding the earliest use of stone in construction in Egypt. In the upland desert areas, readily available fieldstones may have been used for construction from a very early date. Fieldstones were not readily available in the Nile Valley, however, which was dominated by the alluvial soils of the floodplain. Here, construction initially focused on more readily available materials such as timber, reeds, and suitable soils for manufacturing mudbricks. The ancient predynastic regional capital of Hierakonpolis (fig. 11.1), known to the ancient Egyptians as Nekhen, was in almost continuous use from about 4500 BC and likely reached its peak in about 3500 BC by expanding out of the Nile floodplain and inland, along a major wadi. Many years of careful excavations within the largely undisturbed wadi areas have revealed the remarkably well-preserved remains of a major predynastic settlement. Many of these early buildings consisted of a framework constructed from bundles of reeds, with thatched walls, and roofs that were plastered with mud. For larger buildings, timber, including the trunks of acacia trees, was used for the main supports. Thatched buildings seem to have been the predominant form in Egypt until the start of the Early Dynastic Period (fig. 11.3), and although largely replaced by mudbrick, thatch continued in use for certain applications throughout the pharaonic era.

The importance of sun-dried mudbricks in the development of construction in Egypt and other early cultures is often overlooked. Mudbricks provided a dependable and efficient method of construction that, in arid climates such as those that existed throughout most of the pharaonic era, produced remarkably robust and durable structures. Examples of the use of mudbrick from the Predynastic Period are known and include both settlements, such

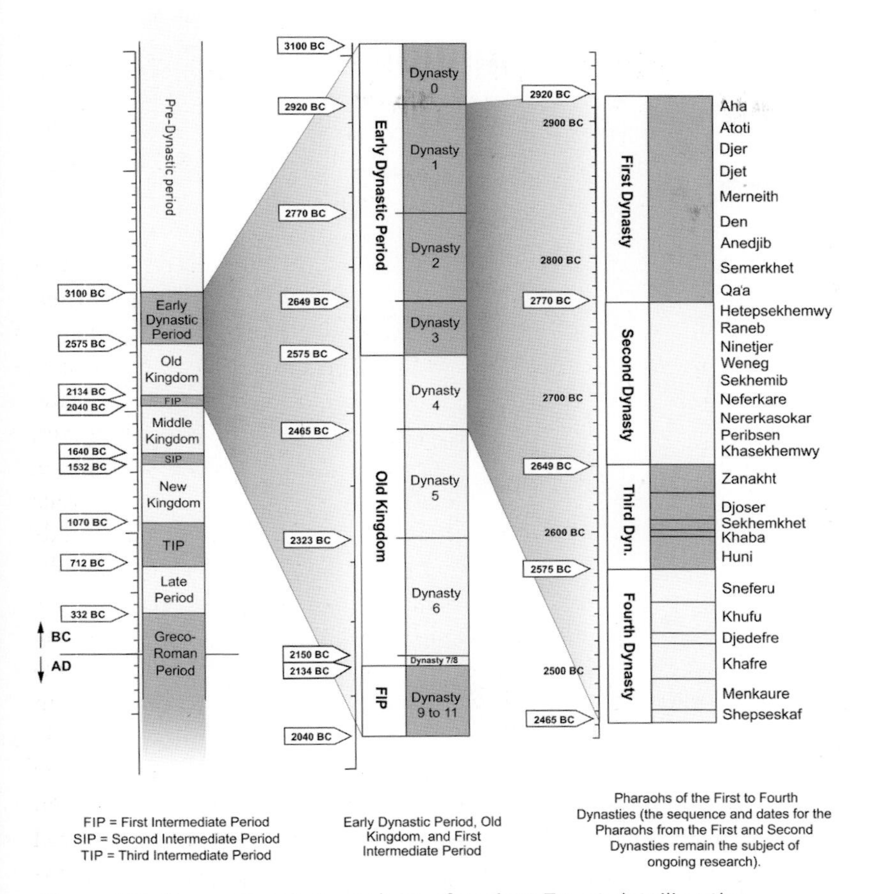

Figure 11.3. Conventional chronology of ancient Egypt detailing the Early Dynastic Period, the Old Kingdom, and the pharaohs of the First to the Fourth dynasties.

as at Tell al-Iswid in the Delta, and burials, such as the predynastic royal tombs at Umm al-Qa'ab, Abydos (fig. 11.1). It was not until the Early Dynastic Period, however, that mudbrick became the primary construction material in Egypt. Used for houses, temples, and defensive city walls, mudbricks became an increasingly useful resource in the developing urban centers. Outside of the main settlements, mudbrick was used extensively in tomb building and for the construction of large funerary enclosures at Abydos and

Figure 11.4. The remains of the "fort" of Hierakonpolis, with mudbrick walls in some areas still standing close to their original height of 10m. Attributed to the late Second Dynasty reign of Khasekhemwy, this is one of the oldest freestanding mudbrick structures in the world. Rather than a defensive structure, it is considered to have been an important element of the pharaoh's mortuary complex.

Hierakonpolis (fig. 11.4). We should not, however, regard mudbrick construction as a precursor to building in stone. In a number of the royal pyramids of the Middle Kingdom, mudbrick was used for the bulk of the structure, while stone was used only for the outer casing. As we saw at the Greco-Roman port city of Dimeh al-Siba in Faiyum (fig. 9.10), mudbricks continued in use throughout the pharaonic era, with stone used only for the most important or prestigious buildings.

Early Experiments with Stone in the Nile Valley

Possibly the earliest large-scale stone artifacts discovered in the Nile Valley are a group of three broken limestone statues that were excavated in the late 1800s at Coptos on the Qena Bend (fig. 11.1). The bodies of these statues were carved from single blocks of limestone and are essentially flattened cylinders with the main features, such as the arms and a girdle around the waist, carved rather figuratively. On the basis of other finds that were recovered, it is likely that the statues had bald, bearded stone heads and would have stood about 4 meters tall. As discussed by Barry Kemp (see Further Reading), the statues are considered to be representations of a fertility god (possibly Min), and it strikes me that some of their main features, such as the right arm held rigidly at the side and the left leg appearing to be placed slightly ahead of the right, are recognizable as distinctly pharaonic. These statues have, however, proved very difficult to date securely, mainly due to a number of typically predynastic or very Early Dynastic features (fig. 11.3). These include the manner in which the knees are represented as pairs of triangles, and motifs carved into the flanks of the statues, including territorial symbols, shells, and animals standing on mountains. On the basis of these features, the Coptos Colossi are thought by some researchers to be much older than the first securely dated use of stone in ancient Egypt. Another example of early stone masonry that has not been securely dated is a sloping wall of rough sandstone blocks that was discovered during early excavations in the main settlement area at Hierakonpolis (fig. 11.1). This masonry had been used to encase a large mound of clean desert sand that is thought to have served as the base for an early temple. The date of the temple and the stone used to dress the faces of the mound on which it was built remains an issue of ongoing debate, with current estimates ranging between the late Predynastic Period and the Third Dynasty.

The earliest reliably dated use of stone in construction in ancient Egypt is the slabs of Aswan granite that paved the floor of the subterranean burial chamber in the tomb of Den, one of the longest-reigning pharaohs of the First Dynasty (fig. 11.3). Den's tomb was built at Umm al-Qa'ab, Abydos (fig. 11.1), a site that, except for a brief period in the early Second Dynasty, had been the focus of royal burials in Egypt from the late Predynastic Period until possibly the start of the Third Dynasty. At their most basic, these early royal tombs consisted of a pit excavated in the sand, the sides of which were lined and supported with mudbrick. As the Early Dynastic Period progressed, these tombs increased in both size and complexity, with internal mudbrick walls used to divide the subterranean space into a number of chambers used for the storage of funerary goods. In addition to the stone floor of the burial chamber, another innovation introduced from the reign of Den was the incorporation of a stairway descending into the tomb, which, unlike earlier tombs, allowed the body to be interred even after the structure was substantially complete. In the tomb of Qa'a from the late First Dynasty, the descending stairway was blocked with a stone portcullis, intended as additional protection against tomb robbers (see Reg Clarke in Further Reading). The First Dynasty royal tombs at Umm al-Qa'ab were surrounded by rows of simpler, shallower graves that were also lined with mudbrick. These secondary tombs appear to have been used for the sacrificial burials of members of the royal court, a relatively short-lived practice that was most likely intended to ensure care for the pharaoh in the afterlife. There is little surviving evidence that either the main tomb or any secondary tombs had a substantial superstructure, although each tomb may have been marked with a low mound of earth and each burial provided with up to two stone stelae bearing the name of the occupant of the tomb (fig. 11.5).

Another important Early Dynastic necropolis was located in northern Saqqara, where a series of often very large mastaba tombs were built on the northernmost section of the Saqqara Plateau (fig. 11.6). When compared with the royal tombs of the same period at Umm al-Qa'ab, the mastaba tombs at Saqqara were often much larger and were built with a highly elaborate superstructure. For many years, Egyptologists considered the tombs at Saqqara to be the main royal tombs, built close to the Early Dynastic capital of the Two Lands at Memphis, with the tombs at Umm al-Qa'ab regarded as cenotaphs, built close to the older and as yet undiscovered capital of Thinis. Recent work at Umm al-Qa'ab, however, has confirmed that the tombs at Abydos are indeed the royal tombs of the First and Second Dynasties, and it is now generally accepted that the mastabas at Saqqara belonged to the most prominent members of Early Dynastic society.

The Saqqara tombs were built entirely from mudbrick. Burial apartments were generally placed below ground level and as the tombs developed and became more complex, were accompanied by additional chambers in the superstructure. Timber was also a major component of these tombs and was used to support chamber roofs. The large mudbrick superstructures of these tombs included elaborately niched façades, which in the case of one First Dynasty tomb from the reign of Djet had three hundred bulls' heads modeled in clay and placed around the outside of the tomb. Real bulls' horns were fitted into these model heads to complete this rather

Figure 11.5. The decorated upper element of one of two stone stelae erected to mark the tomb of the First Dynasty queen Merneith in the traditional royal cemetery at Umm al-Qa'ab, Abydos. Merneith may have ruled Early Dynastic Egypt in her own right between the reigns of Djer and Den.

dramatic effect. Many of the tombs included geometric patterns painted both inside and outside, and in the Third Dynasty tomb of Hesy-Re, intricately carved, imported cedarwood panels were used to line the chambers of the tomb. As we saw at Umm al-Qa'ab, the Saqqara burials included occasional stone elements, the earliest of which may be roofing slabs used in a tomb built during the early First Dynasty reign of Djer (fig. 11.3). Possibly from around the same period is a limestone lintel decorated with lions from a tomb that has been attributed to Merneith, a queen of Djer and mother of Den. Of more secure date is the tomb of Hemaka, a noble from the reign of Den, which included a number of stone portcullises within the internal passages of the tomb.

The tombs from Abydos and Saqqara represent the burials of the most elite elements of early pharaonic society; however, cemeteries of lower-ranking Early Dynastic Egyptians have also been

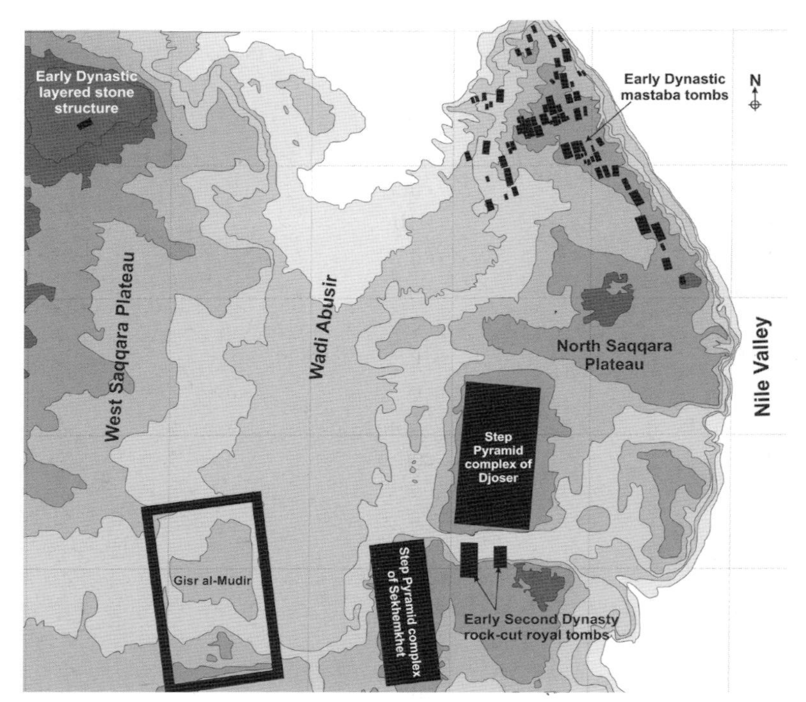

Figure 11.6. Plan of north Saqqara showing the area of Early Dynastic mastabas in the north and the principal stone-built Early Dynastic monuments that are mentioned in the text.

identified. Of these, perhaps the largest and best understood is at Helwan, a southern suburb of modern Cairo (fig. 11.1). The Helwan site was in use from the late Predynastic Period through to the Old Kingdom and beyond, and a number of the tombs excavated at the site included stone elements in their construction. The earliest use of stone at Helwan is thought to date from the later part of the First Dynasty (again from around the reign of Den), when a number of tombs, constructed as open pits in the gravel soils of the area, were built with walls and floors lined with slabs of locally sourced limestone. Several Second Dynasty tombs have also been identified that include similar stone elements, but exhibit a number of more advanced construction techniques, such as coursing of the masonry. A single stone-built structure at Helwan, interpreted as a ruined mastaba tomb, has been dated to the late Third Dynasty (fig. 11.3).

The evidence from sites such as Abydos and Saqqara suggests that the use of stone in construction began in the First Dynasty, certainly by the reign of Den but possibly a little earlier. It is important, however, that we recognize these examples as outliers. None of these early examples of the use of stone can be interpreted as the beginning of a sustained tradition in the use of stone masonry in ancient Egypt. Although there appears to be more continuity at Helwan, with stone elements found in tombs from all three dynasties of the Early Dynastic Period, these represent fewer than twelve tombs in a cemetery of more than 10,000 burials. Numerically at least, the use of stone at Helwan appears to be the exception rather than the rule. The generally accepted evidence suggests, therefore, that stone was used only occasionally in construction in ancient Egypt during the Early Dynastic Period.

Early Stone Buildings

An entry on the Palermo Stone (chapter 6) refers to a pharaoh of the late Second Dynasty who built a structure in stone. Given that the Palermo Stone is incomplete, we do not know precisely who this celebrated builder in stone was; however, it is possible that this text is referring to the use of stone in the burial chamber of the last pharaoh of the Second Dynasty, Khasekhemwy (fig. 11.3). Although Khasekhemwy's tomb at Umm al-Qa'ab was built largely in mudbrick, the large central burial chamber was constructed entirely from quarried limestone. Another candidate for the early stone building referred to on the Palermo Stone is the Gisr al-Mudir at Saqqara (fig. 11.6). The Gisr al-Mudir is a huge walled enclosure the majority of which was built from limestone masonry. At over 600 meters long and 400 meters wide, the walls of the Gisr enclose an area about twice that of the enclosure that surrounds the nearby Step Pyramid of Djoser (see below). Because the walls of the Gisr have been robbed away and its low-lying remains are largely obscured by wind-blown sand, the Gisr is not a particularly prominent feature of the modern

Saqqara landscape; however, when complete, its towering stone walls must have stood in excess of 5 meters high. Many of the features of the masonry used to build the walls of the Gisr al-Mudir, including the irregular size and shape of individual blocks, the often haphazard placing of the masonry, and the simple manner in which the corners of the enclosure were constructed, contrast greatly with the more refined masonry techniques used in the early Third Dynasty Step Pyramid of Djoser, leading to the conclusion that the Gisr al-Mudir may therefore be an earlier, Second Dynasty structure, perhaps the monument referred to on the Palermo Stone.

In terms of construction in stone, the early Third Dynasty marks an extraordinary period in the pharaonic era. Built during the reign of the second pharaoh of the Third Dynasty (fig. 11.3), Djoser's Step Pyramid is quite unlike anything that went before. Technological progress in Early Dynastic Egypt was such that by the reign of Djoser, mastabas of the most prominent people in Egypt had evolved into huge mudbrick structures such as those at Beit Khallaf, north of Abydos (fig. 11.1), the largest of which was 45 meters by 85 meters in plan and 8 meters high. The first ambitious move adopted for Djoser's tomb at Saqqara was to take the basic concept of the huge mudbrick mastaba and translate it into a structure built entirely from stone. The second audacious move was to stack a series of ever-smaller stone mastabas, one on top of the other, to produce Egypt's first pyramidal structure.

The 62-meter-high Step Pyramid (plate 23) is a solid mass of masonry, with its burial chamber and an associated maze of shafts and passages all running through the bedrock underlying the pyramid. Although Djoser's pyramid is remarkable, it is the complex of structures that surround the pyramid that is perhaps the more astonishing. The 10-meter-high wall that defines the Step Pyramid enclosure is built with a core of rough limestone masonry that was cased with very finely worked, high-quality limestone blocks that were carefully laid in regular courses, often using a broken-bond pattern of the type that is still used in modern masonry. Each block had meticulously shaped edges and corners and was fitted to its neighboring block with absolute precision. This stonework was far more advanced than anything used in the Gisr al-Mudir. Both the inside and outside faces of the enclosure wall were built with a detailed pattern that was a development of the niched façade that had decorated the external faces of the Early Dynastic mudbrick mastabas in the north of Saqqara (fig. 11.6); however, this niched pattern was interrupted at locations around the wall by a series of outward-projecting buttresses. One of these buttresses, near the southeast corner of the enclosure, has a tall, narrow entranceway that runs through the thickness of the wall and provides the only access into the entire enclosure.

Once inside, the sheer variety of stone-built features is a testament to the ancient designers and skilled builders of this incredible monument. Just beyond the entrance is a long, roofed colonnade, and flanking the two sides of the corridor are chambers, the walls of which end in columns carved from blocks of limestone to resemble bunches of reeds that had been tied together. The limestone slabs originally used to provide a roof over this colonnade were carved to imitate wooden logs. In these columns and roof slabs, as in so many other features within the Step Pyramid enclosure, we see building elements that for centuries had been constructed from perishable materials such as reed or timber, reproduced for the first time in stone. Djoser's Step Pyramid enclosure was built to last for eternity. The colonnade opens up to a pavilion-style court, with the Step Pyramid at the northern end. To the south, east, and west of the court are additional masonry buildings, each of which was probably intended to serve a specific role for the pharaoh in the afterlife. One building has a frieze of striking cobras running along the top of the wall; some buildings have arched roofs with attached limestone columns fashioned to resemble bundles of reeds, and others have flat roofs, all of which are rendered in the most amazing stonework. These, however, are not functional buildings. They were built as a solid mass of masonry with no usable internal space, perhaps intended for use only by the spirit of the dead pharaoh. As discussed by Jean-Philippe Lauer (see Further Reading), the numerous buildings within the Step Pyramid enclosure have all been painstakingly reconstructed from ancient blocks of masonry that have been uncovered over the past hundred years, as Egyptologists have carefully excavated the site. A large area north of the pyramid has yet to be thoroughly cleared, and it remains to be seen what other incredible structures will be found there.

There is one feature of the Step Pyramid enclosure that is seldom discussed, yet from the perspective of this book on the geology and landscape of ancient Egypt, it is perhaps one of the most remarkable. We spoke briefly in chapter 10 about the Tura quarries and the white limestone used to case the pyramids and other important pharaonic buildings. The casing used in the Step Pyramid enclosure has developed a rich brown patina that is very different from that which can be seen elsewhere on surviving sections of Tura limestone masonry. Studies by Dietrich and Rosemarie Klemm (see Further Reading) have confirmed that the casing stone used in the Step Pyramid enclosure was not quarried at Tura, and yet the actual location of the quarry has not been found. It is possible that this unique stone was obtained from an outcrop closer to Saqqara, an outcrop that was completely removed by these early quarrying activities. If so, this is a rare early example of the ancient Egyptians permanently and irreversibly modifying the landscape around them. As the pharaonic era unfolded, there were many more examples in which the landscape of the Nile Valley and the fringes of the desert were modified, but here at Saqqara we see, for possibly the first time, the cultures of the Nile Valley developing the courage to shape the physical world in which they lived.

As we saw with the early experimentation with stone at sites such as Umm al-Qa'ab and Helwan, once again the technological advances evident in the Step Pyramid represent something of a false start. Djoser's successor, Sekhemkhet, began the construction of a smaller step pyramid complex in the area between the Gisr al-Mudir and Djoser's Step Pyramid at Saqqara (fig. 11.6), but Sekhemkhet's reign seems to have been short and construction of his tomb complex had barely begun at the time of his death. The pyramid of the subsequent pharaoh, Khaba (fig. 11.3), was built at Zawiyat al-Aryan, some distance northwest of Saqqara, and had a layout that was very different from Djoser's complex. Controversy surrounds the pyramid of Khaba's successor, Huni, who, as I explored in a paper published

Figure 11.7. The two enormous stone pyramids built by Sneferu at Dahshur. The pyramid at the left was probably built first and is called the Bent Pyramid because of the change in angle about halfway up. Some researchers see the change of angle as deliberate; however, it seems more likely that it was introduced during construction to address potential faults in the growing pyramid. It is thought that because of these faults, Sneferu had a second huge stone pyramid built nearby, the Red Pyramid (right).

in 2015 (see Further Reading), appears to have begun the construction of a step pyramid at Meidum, far to the south of Saqqara (fig. 11.1). Huni's unfinished step pyramid may have been completed by his successor Sneferu, the first pharaoh of the Fourth Dynasty, but the extent to which Sneferu was involved at Meidum remains the subject of a great deal of ongoing debate. From the uncertainty of the reign of Huni emerges the great age of pyramid building in ancient Egypt, heralded by Sneferu, who, for reasons not yet fully understood, built two enormous stone pyramids at Dahshur (fig. 11.7). In the following reigns we see the construction of the pyramids of Khufu, Khafre, and Menkaure at Giza (plate 24), together with a succession of other, often much smaller pyramids built across a range of sites throughout the Old and Middle Kingdoms (fig. 1.6). As discussed by Mark Lehner and Ahmed Fakhry in their respective publications (see Further Reading), much has been written about the pyramids of Egypt, but never again do we see a pyramid complex as startling, as ambitious, or as revolutionary as that built for Djoser in the early Third Dynasty.

Building the Pyramids

One of the most enduring mysteries of ancient Egypt is how the great pyramids were built. At the root of this problem lies the fact that pyramid-building required millions of tons of masonry to be quarried, transported, and placed, often to great heights and often with tremendous precision and accuracy. A variety of theories have been suggested to explain how the pyramids were built, some of which are less well argued or have been discredited entirely as additional evidence has come to light. One such discredited theory suggested that rather than build the pyramids from quarried blocks of stone, a concrete-like material was used. This theory was not widely accepted when it was initially presented and has recently received a further setback by discoveries at Wadi al-Jarf on the Red Sea coast (fig. 11.1). The Wadi al-Jarf site appears to have been an Old Kingdom port, and among other remarkable finds, Pierre Tallet and his colleagues have recovered the remains of a series of papyrus scrolls, the oldest so far known from ancient Egypt (see Further Reading). From among the papyrus fragments, it has been possible to reconstruct part of the diary of a man named Merer. Merer's diary dates from the later part of the reign of Khufu, the builder of the Great Pyramid at Giza, and records numerous sailing expeditions, during which Merer and his crew transferred blocks of casing stone from quarries at Tura across the Nile, and on to the Giza Plateau. Clearly, there would have been no need for Merer to make these sailing trips if the ancient builders at Giza had been using a form of concrete to build the pyramids.

Recent excavations at Giza have found both the camp and the cemetery of the workforce that built the pyramids, and these discoveries have established beyond doubt that, rather than slaves, the workforce were well fed, provided with medical care, and from texts left in their tombs, appear to have been very proud of their involvement in these prestigious projects. As a result of the physical demands of their job, however, it appears that they died at a relatively young age compared with the rest of ancient Egypt's population.

Although evidence for the workforce has been found, we have yet to find clear evidence for the actual construction methods used to build the pyramids. The use of external ramps to raise building materials to height is well attested at ancient sites throughout Egypt, but given the enormous scale of the pyramids, many researchers continue to argue against the use of external ramps in their construction. One of the main objections that surfaces repeatedly is the sheer volume of material that would have been required to construct effective ramps for use in pyramid building. Not only did material need to be placed initially to build the ramp, it also needed to be taken away when the pyramid was finished. As the pyramids themselves clearly demonstrate, however, the pharaonic Egyptians were quite willing and able to move and place enormous volumes of material. Furthermore, external ramps offer a number of advantages that are often overlooked. These advantages include the ready availability of the material used for their construction (most probably sand and mudbrick) and the fact that ramps are relatively simple structures, the shape of which could be readily modified to adapt to changing requirements as construction progressed.

Bob Brier and Jean-Pierre Houdin (see Further Reading) have proposed an alternative theory which suggests that internal ramps may have been used, built within the structure of the pyramid. If internal ramps were used, evidence should remain in the form of voids within the pyramid structure, and a number of programs of nondestructive testing have been undertaken to locate such evidence. Among the early findings of these investigations is the potential presence of a large and previously unsuspected cavity high in the superstructure of the Great Pyramid, along with evidence for a number of other potential voids that may possibly be associated with internal ramps. As with the results of all such studies, however, care needs to be used when interpreting this data. For example, it has been recognized for some time that the masonry of the pyramids was not worked to the same consistent standards throughout, and although there are many examples in which masonry has been carefully worked, there are also examples, particularly in the core of the pyramids, in which the masonry appears to have been quite roughly finished. It is therefore possible that, rather than open spaces within the structure of the pyramids, the anomalies that have been identified during the recent nondestructive testing may be areas of less carefully constructed masonry. Given that the precise nature of these anomalies cannot be established until physical investigation has been undertaken, we must await the outcome of further research before we can reach any firm conclusions.

Although mechanical lifting devices were mentioned by Herodotus and have been favored by a number of theorists ever since, their main drawback is likely to have been the limited loads that such devices could have tolerated, due largely to the limitations of the materials available to the ancient builders. The weakest parts of any lifting device are the points at which loads are focused, such as pivots or axles. As we saw in chapter 10, the only metals available in the Old Kingdom were copper and copper alloys. If these metals were used for axles and pulleys, for example, it is unlikely that they would have had sufficient strength to handle the loads that would have been imposed, particularly when lifting the multi-ton blocks of masonry used during the main phase of pyramid building. When the relative weakness of the materials available is considered alongside the number of individual lifting devices that would have been required to raise the millions of large stones used in the construction of the Great Pyramid, for example, it seems highly unlikely that mechanical lifting devices could have played anything but a minor role in the construction of Egypt's major pyramids.

Clearly, a great deal of further study is required if we are to have any greater understanding of the methods used to build the Egyptian pyramids. Whatever the outcome of these future studies, I feel there are a number of construction-related mysteries that will endure. Perhaps the greatest of these is the precision with which many of the pyramids were built. For example, the four sides of the Great Pyramid of Khufu are nearly identical in length and are nearly perfectly aligned with the cardinal points of the compass. Although theories have been presented to explain how these remarkable levels of precision may have been achieved, they all include methods that will have been prone to some degree of error, however small. It's likely that the significance of these inherent errors increased with the scale of ancient Egypt's building projects. Given these issues, I still struggle to comprehend how the ancient Egyptians of 4,500 years ago achieved such remarkable levels of precision on such a vast scale.

The Valley of the Kings

The pyramids of the Old and Middle Kingdom were extraordinary monuments that dominated the landscape and served to proclaim the absolute power of the pharaohs who were buried within them. Despite the many safeguards that were put in place to keep the burials secure, however, most of Egypt's pyramids were robbed, probably in antiquity. In the early New Kingdom, the need for the pharaoh's remains to lie undisturbed for eternity led to a dramatic change in royal funerary architecture in Egypt. Gone was the ostentation of the highly visible pyramid, to be replaced with far more secretive tombs that, as discussed in chapter 8, were hidden in a cluster of otherwise unremarkable wadis opposite the royal capital at Thebes, behind the western cliffs of the Eonile Canyon. When selecting these wadis for royal burials, the ancient Egyptians did not completely sever the link with the pyramid as a key element of the royal necropolis. Dominating the western Theban skyline and standing over this group of sacred wadis is a natural pyramid, the triangular peak known as al-Qurn that is visible on the left in plate 2.

The most well-known of these wadis is the Valley of the Kings, a sacred valley in which many of the pharaohs of New Kingdom Egypt were buried. The hidden tombs of the Valley of the Kings descended from the walls and floor of the wadi, often for tens of meters through the bedrock, to a series of chambers in which the burial and burial goods were placed. The limestone beds into which the tombs were cut date largely from the early Eocene (fig. 2.1) and are soft strata that would have been relatively easy to work. Underlying the Eocene rocks at or below the base of the valley, the older Esna Shales were encountered in many of the tomb passages and chambers. These fine-grained Paleocene strata (chapter 5) have a tendency to swell, particularly when in contact with water, and would have presented a number of difficulties for the ancient tomb builders, as they do for modern conservators.

Many of the tombs in the Valley of the Kings, including a number used for nonroyal burials, were relatively simple, with a short corridor leading to a single chamber. From the outset, however, the pharaonic tombs were far more elaborate structures, and no two are the same. The main entrance passage of royal tombs often consists of several straight sections, with each section leading to a small antechamber. In most cases these antechambers herald a change in direction or angle of descent for the subsequent section of passage. Uniquely, the passages in the earliest pharaonic tomb (KV20, the tomb of Thutmose I and Hatshepsut—see Kent Weeks in Further Reading) are long and curved, providing the tomb with a U-shaped arrangement as it descends nearly 100 meters through the limestone bedrock. At the lower end of the descending passage, tombs opened out into a series of chambers that formed the main burial apartments. These chambers are often unexpectedly large, with the ceilings supported by columns of bedrock that were deliberately left in place as the chamber was excavated around them.

Today these tombs are equipped with ventilation and electric lights, and it is easy to overlook the difficulties that the tomb builders had to overcome. With only naked flames for light, no ventilation, and a confined subterranean environment thick with dust from cutting the limestone, the builders had to create these awe-inspiring spaces from scratch. It is likely that each burial chamber was begun at ceiling level, initially by working a narrow passage upward from the end of the entrance corridor, before widening the excavation to create a wide, flat cavity covering the approximate area that was required. It would have been important at that stage to make sure the size and shape of this initial excavation was sufficient and that the corners of the ceiling were neatly squared, because once excavation began to work down toward the floor of the chamber, the ceiling would have quickly become out of reach. Any columns required to support the ceiling would need to have been identified at this early stage. Once the excavations were complete, we can only assume that an internal scaffold was erected to allow the space to be appropriately decorated. Given the generally poor quality of the limestone from which the tombs were cut, in most instances the chambers were plastered before being decorated in vivid colors. The ceilings were painted with a dark blue background and golden-yellow stars representing the night sky. On the walls, scenes of the pharaoh and of the gods were surrounded by hieroglyphic texts intended to assist the deceased ruler during his passage through the underworld. In many cases, these vivid pigments have survived the centuries remarkably well.

Of more than sixty tombs in the Valley of the Kings, two deserve special mention. KV5 is quite unlike any of the surrounding tombs. A short descending passageway leads to a pair of antechambers and then to a large, sixteen-pillared hall. From this main hall, although corridors run in many directions, they preserve a rigorous orthogonal layout, and along each of these main corridors, countless small side chambers were excavated, all of which are thought to be burial chambers for the children of Ramesses II: Ramesses the Great, who probably lived into his nineties. Despite many years of painstaking work, KV5 has still not been fully excavated and its full extent remains unknown. The other tomb that simply cannot go unmentioned is KV62, the tomb of Tutankhamun. The tale of the discovery of this tomb by Howard Carter in 1922 has been told frequently (fig. 1.7). Tutankhamun's relatively short reign and early death at about nineteen years of age may explain why his tomb is unusually small and appears to have been constructed hurriedly, possibly by adapting an already existing tomb. Recently, Nicholas Reeves has raised the intriguing possibility that there may be additional elements to the tomb that remain undiscovered, or that a more extensive tomb of one of Tutankhamun's royal predecessors may lie beyond the chambers that have been identified so far (see Further Reading). Attempts to address these theories without causing damage to the tomb and its ancient decoration have led to a number of phases of investigation using ground-penetrating radar along with other technologies, but so far, the findings have been inconclusive.

Despite the precautions that were taken to maintain the secrecy of the Valley of the Kings and the security measures that were put in place to protect individual tombs, the vast majority of the burials were plundered in ancient times, most likely by conspirators among the tomb builders or the guardians that were employed to protect the tombs. Although there had been some minor disturbance of Tutankhamun's tomb in the years immediately following the king's death, it was thought until recently that this relatively small and simple burial had survived because the short-lived king had been quickly forgotten, perhaps because, like his "heretic" father Akhenaten, his memory had been deliberately erased soon after his reign had ended. The fate of Tutankhamun's tomb, however, serves as another vivid example of the important contribution that geology can play in the field of archaeology. A number of the tombs in the Valley of the Kings have been found with passages and chambers choked by flood debris, the result of flash floods that occasionally ripped through the wadi. Although efforts were undertaken in the later New Kingdom to protect the valley from further flooding, at the time of Tutankhamun's burial these flash floods went unchecked. As Steve Cross has identified, the entrance to Tutankhamun's tomb lies near the bottom of the Valley of the Kings, and Cross's careful interpretation of the ground conditions has shown that within perhaps a few years of the king's burial, the entrance to the tomb was lost under flood debris (see Further Reading). Hidden from view, the tomb lay undisturbed for over 3,000 years before Carter's momentous discovery. We can therefore add the survival of the tomb of Tutankhamun and its magnificent treasures to the list of geological "events" that have greatly influenced our understanding of the pharaonic culture of the Nile Valley.

Decorating Hard Stones

Finally, I would like to discuss another enduring mystery: how the ancient Egyptians were able to carve elaborate decoration, including detailed hieroglyphic texts, with such precision into the hardest of stones. Denys Stocks (whom we met in chapter 10) has demonstrated that tubular copper drills of various diameters can be used to reproduce many of the forms of decoration that the ancient Egyptians applied to hard stones (see Further Reading). His experiments have demonstrated that by rotating these drills using simple wooden bows and using quartz sand as a cutting agent, granite and other stones can be drilled quite effectively. Many examples of masonry that have clearly been worked by tubular drills have been found in Egypt, and a number of tomb scenes have been identified that appear to show these drills in use. I would, however, like to insert a note of caution and remind ourselves that despite understanding some of the fundamentals of ancient technologies such as stoneworking, we have yet to establish many of the details. Plate 25 is a photograph of a granite lintel from Tanis in the Delta (fig. 11.1). This lintel is carved with a cartouche (the oval shape that held the name of the pharaoh) of Ramesses the Great. Ramesses' name has been cut deeply into the granite in the precise manner that was typical of this period. The carving, however, is incomplete. To the right of the cartouche, we see a representation of a bee, one of the classic components of the pharaoh's title. Although largely finished, the front of the bee and the leading wing are clearly incomplete and the surrounding surface of the lintel has a rough, partially worked appearance. To the right of the bee, other elements of the royal title, including the sedge plant and the ankh, are far from complete, appearing just to have been pecked lightly into the surface of the granite.

This lintel from Tanis was clearly never finished, but it is often incomplete examples such as this that are invaluable to modern researchers, in this case revealing elements of the process involved in the carving of hieroglyphs into granite. Although it is always difficult to assume too much from a single example, there is no evidence here for the use of drills in carving hieroglyphs. When each drill hole has been completed, the drilled-out core needs to be broken off, leaving a characteristic stump at its base. Similarly, closely spaced, overlapping drill holes would leave a series of small cusps that would need to be removed. The unfinished carving from Tanis has neither stumps nor cusps, or any other indication that drills were being used in the decoration of this lintel. Judging from the partly worked symbols to the right of this decoration, it appears that the individual characters of this text were picked out by some form of progressive hammering or chiseling process. Although the carving could have been achieved using hardened chisels made from copper alloys, or small hammer stones made from dolerite or other hard material, these would be far less efficient than the use of drills, and it is difficult to understand how the sharp internal corners at the base of each hieroglyphic character could be achieved using simple hammering methods. Although I am a great supporter of Denys Stocks's research, I cannot explain how this unfinished granite lintel from Tanis was carved.

The pharaonic era of Egypt is rightly known for it its incredible stone structures, including not only the pyramids, but also great temples, such as those at Karnak, Luxor, and Abu Simbel. These later buildings owed a great deal to the pioneering use of stone in structures of the Early Dynastic Period and Old Kingdom, and without the early experiments in stone discussed in this chapter, pharaonic architecture would not have developed to the extent for which it is justly renowned. If it were not for the early experiments in stone such as those as Abydos, Saqqara, and Helwan, it is unlikely that the banks of the Nile would be dominated by the magnificent structures that we marvel at today, over 2,000 years after they were built. As the granite lintel from Tanis demonstrates and as we will explore further in the next chapter, however, we are still a long way from understanding many of the secrets of ancient Egypt.

CHAPTER 12
THE INTERACTION OF MONUMENTS AND LANDSCAPE

In the last chapter we explored how the use of stone in building may have developed in ancient Egypt. The earliest known and reliably dated use of stone appears at some point around the reign of Den in the First Dynasty, with stone elements continuing to be used in a relatively limited manner throughout the Early Dynastic Period, at sites such as Abydos, Saqqara, and Helwan. This development appears to have culminated in the late Second Dynasty with the first major use of stone at the Gisr al-Mudir. If, like some researchers, you are not willing to accept a late Second Dynasty date for the Gisr, there can be no doubt that shortly after, the construction of the early Third Dynasty Step Pyramid of Djoser marks a significant landmark in the use of stone in construction in ancient Egypt.

Like the Coptos Colossi and the sloping sandstone wall at Hierakonpolis, the Gisr al-Mudir illustrates some of the current difficulties that exist for understanding the development of stone in construction in ancient Egypt. Many features of the Gisr appear crude, suggesting that this was perhaps the first time the masons were working with stone on a large-scale construction project. However, the Gisr al-Mudir has not been securely dated because only limited excavation and study has been undertaken. The absence of definitive dating evidence for the Coptos Colossi or the sandstone masonry at Hierakonpolis has arisen because these examples were found during the early phases of exploration in Egypt, which tended to focus on the artifact, with little or no attention paid to the context in which each artifact was found. As a result of this early approach to excavation, much critical dating evidence for these and many other monuments and artifacts has been lost. These are not criticisms. Prior to the late 1800s, excavations in Egypt were undertaken primarily to recover exhibits for museums or private collections, and it required the enlightened intervention of individuals such as Sir William Matthew Flinders Petrie before the discipline of Egyptology could develop. Likewise, the limited scale of investigation at monuments such as the Gisr al-Mudir is simply an indication of the enormous archaeological wealth of Egypt, the study of which will require centuries of concerted effort.

Recent excavations combined with new approaches to the interpretation of a number of more familiar monuments in Egypt, however, suggest that it may already be necessary to revisit some aspects of our understanding of construction in stone. For example, an oval-shaped, stone-built substructure has been identified in Ma'adi (a southern suburb of modern Cairo, see fig. 11.1) that predates the reign of Den by as much as 500 years (see a paper by Luc Watrin and Olivier Blin in conference proceedings edited by Zahi Hawass in Further Reading). Given that this structure was revealed during recent excavations, there seems little reason to question its early date. An unusual layered stone structure has also been found in a relatively remote part of Saqqara, about 2 kilometers to the northwest of Djoser's Step Pyramid (fig. 11.6, top left, and fig. 12.1). Built against the lower flanks of a prominent hill, this structure consists of a masonry platform with associated rock-cut chambers and has been dated to the late Second or early Third Dynasty, on the basis of the style of the masonry used and the relatively crude methods of construction. Although often interpreted as an early cult temple, the exact purpose of the monument remains unknown (see a paper by Sakuji Yoshimura and Masanori Saito, also in the proceedings edited by Zahi Hawass in Further Reading).

Figure 12.1. The unusual layered stone structure to the northwest of the Saqqara necropolis. Although the purpose of this structure has not been established with any certainty, it has been provisionally dated to the Second or Third Dynasty.

From the same general period in the late Second or early Third Dynasty is one of the oldest dams in the world, the Sadd al-Kafara dam in Wadi al-Garawi, to the southeast of Helwan (fig. 11.1). First identified during the late 1800s, it is likely that this dam was built to protect downstream sections of the wadi from the effects of unusually heavy rains that are known to have occurred in Egypt at about that time. Despite its age, the dam was surprisingly sophisticated. Like many modern dams, it consisted of an earth core surrounded by dressed masonry buttresses that were constructed from blocks of limestone, quarried from the nearby walls of the wadi (see the entry "Wadi Garawi Dam" in Kathryn Bard's *Encyclopaedia of the Archaeology of Ancient Egypt* in Further Reading). The dam appears to have been under construction for ten or twelve years, and if completed, would have been some 113 meters long, 14 meters tall, and 98 meters wide at the base. The project was abandoned after part of the central section of the dam was washed away by one of the destructive floods that, ironically, the dam was intended to offer protection against. One of the most remarkable features of this ancient structure is that, unlike most early Egyptian stone monuments, this is not a tomb or temple. The dam was built to serve a more practical purpose and its remains stand as testament to the immense ingenuity of the people of ancient Egypt, who, despite regarding the Red Land in which the dam was built as the realm of unpredictable and powerful gods, were willing to modify the landscape on a significant scale for the benefit of their society.

Significant modification of the natural landscape was also required to construct one of the most instantly recognizable monuments from ancient Egypt, the Great Sphinx of Giza (plate 6). The Great Sphinx was not built from masonry but was carved out of a layered sequence of limestone strata that originally formed a low hill on the eastern edge of the Giza Plateau. It is generally believed that the Great Sphinx was carved in the Fourth Dynasty as part of the pyramid complex of Khafre (figs. 10.7 and 11.3); however, this is based on largely circumstantial evidence. As the only large-scale, freestanding sculpture carved from bedrock that is known from the entire pharaonic era, it has proved difficult to establish a definitive date for the construction of this great, human-headed lion. Recently, a number of researchers have remarked on the unusual condition of the limestone exposed at the Sphinx. By examining the weathering and erosion of these limestones and the impact that quarrying at Giza had on the pattern of rainfall runoff across the site, I have concluded that the origins of the Sphinx lie in the Early Dynastic Period (see my 2001 paper in Further Reading). These conclusions require there to have been development at Giza before the first pyramid was built there, and although this is at odds with the generally accepted history of the site, it is consistent with a range of often overlooked evidence for early development at Giza that has been presented by mainstream Egyptologists (for example, see the separate papers by Peter Der Manuelian and Maira Torcia in Further Reading).

Interactions with the Landscape

The modifications to the landscape required to construct the Great Sphinx or to build the Sadd al-Kafara dam represent only one aspect of the interaction of landscape with early construction in Egypt. Another aspect that needs to be considered is the manner in which the landscape itself influenced ancient development, and I have spent several years studying this interaction at Saqqara, one of the most important sites in ancient Egypt. Saqqara (fig. 11.6) served as the principal necropolis and cult center for Memphis throughout the pharaonic era, and the site was arguably at its most important during the Early Dynastic Period and Old Kingdom, when a number of royal pyramids were built there (chapter 11). As the Saqqara necropolis developed, it became increasingly likely that the location selected for a particular monument would be influenced by the presence of existing pyramids, tombs, and temples. Go back far enough in time, however, and Saqqara would have been free from such constraints, suggesting that the location selected for the earliest monuments was influenced by other factors. I consider that landscape was the most significant of these factors and had both a practical and a theological influence on the earliest phases of development of the site.

The practical influence of landscape on construction can be readily addressed. Ancient monuments needed a sufficiently large, relatively flat area, free from poor ground conditions and close to an available source of suitable building material. Most areas of the elevated limestone plateau at Saqqara meet these basic requirements, and so there is little reason to consider that these practical issues had a significant influence over where a particular early monument was built. There is, however, a growing body of evidence to suggest that theology had a considerable influence on early site selection. Many of the monuments at Saqqara tend to have an east–west orientation, and this has led to the generally held view that the development of the site focused on the Nile Valley and the capital city of Memphis that lay nearby. As I discussed in a paper published in 2017, however (see Further Reading), some of the earliest phases of development at Saqqara appear to have been influenced by another landscape feature, the north–south aligned Wadi Abusir. The Wadi Abusir separates the main pyramid plateau at Saqqara (the North Saqqara Plateau, fig. 11.6) from other elevated areas to the west, and although the wadi is unlikely to have seen any significant surface water for thousands of years, even today a limited area of scrub vegetation at its northern end adds a little green to the otherwise barren landscape (plate 26). As we have seen throughout this book, Egypt experienced generally less arid conditions in the earliest parts of the pharaonic era, when, as described by Karl Butzer, wadi vegetation would have been more extensive than it is today (see Further Reading). During the earliest phases of development at Saqqara, therefore, vegetation within the Wadi Abusir would have been more abundant, extending to the south along the wadi's main axis (fig. 11.6). In my research, I concluded that this swath of vegetation and the associated dark-colored soils were likely to have been regarded by the Early Dynastic Egyptians as part of *kemet*, an arm of the Black Land that extended out of the Nile Valley and deep into the surrounding desert. As such, the vegetated Wadi Abusir provided an area of relative safety, a route that allowed the living to travel into the Red Land to perform burials or associated rituals without having to venture into the realm of the dead and its spiritual chaos and disorder.

The Development of the Saqqara Necropolis

The Early Dynastic mastaba tombs in the north of the necropolis (fig. 11.6) are among the oldest phases of development at Saqqara. In the First Dynasty, these elaborate mudbrick tombs were strung out along the eastern edge of the plateau, leaving little doubt that they were intended to be seen from the Nile Valley. During the Second and Third Dynasties, however, the early necropolis moved away from the edge of the escarpment, extending westward across the plateau toward the Wadi Abusir. Research by Elaine Sullivan, who has undertaken 3D computer modeling as part of her studies on the wider landscape of the Memphis region, has confirmed that the Second and Third Dynasty mastabas in the north of Saqqara were not visible from the Nile Valley (see Further Reading). Although the westward shift of these tombs may be seen as inevitable, given that the earlier First Dynasty tombs occupied the eastern escarpment edge, if visibility from the Nile Valley had continued to be a key consideration, this could have been accomplished by extending the necropolis further south along the edge of the plateau or, as happened in a number of examples, reusing the site of older tombs. After an initial focus on the Nile Valley, therefore, as the Early Dynastic Period progressed, the Wadi Abusir appears to have taken on greater significance, with visibility from the wadi becoming an increasingly important consideration. As if to confirm the importance of the wadi at this time, in the late Second or early Third Dynasty, the layered stone structure referred to earlier in this chapter (fig. 12.1) was built on another elevated site, directly across the mouth of the wadi (fig. 11.6), far away from the Nile Valley. Having been built against the foot of a hill, this early stone structure offers no visibility to the north or to the west. The structure's southeast-facing façade, however, presents uninterrupted views across the Wadi Abusir and the North Saqqara Plateau beyond.

For reasons that are not fully understood, at the start of the Second Dynasty the focus of royal burials in Egypt shifted from the traditional necropolis at Umm al-Qa'ab, Abydos, to North Saqqara, with two of the first three kings of the Second Dynasty (Hotepsekhemwy and Ninetjer, fig. 11.3) buried in rock-cut tombs in the southeast of the necropolis, as shown in figure 11.6. Evidence for the burial of another early Second Dynasty pharaoh, Raneb, has also been encountered in this area, but to date, his burial place has not been identified. The construction of these early Second Dynasty royal tombs at Saqqara was a great break with tradition. This was the first time that Egyptian monarchs had not been buried at the royal cemetery of Umm al-Qa'ab, and yet despite their potential importance, it is only recently that the tomb of Ninetjer has undergone extensive study. The tomb of Hotepsekhemwy still awaits modern investigation. Rather than select a prominent location, these tombs were built in a relatively unremarkable minor wadi, accessible from the Wadi Abusir. It is not known whether these tombs had any sort of superstructure, but their location suggests that security and isolation had been key factors when choosing where they were to be built. Some researchers have suggested that the Gisr al-Mudir (fig. 11.6), which lies across the Wadi Abusir from the Second Dynasty royal tombs, may be associated with these royal burials. The tombs, however, date from the start of the Second Dynasty, whereas the limited available evidence we have points to the Gisr being built later in the Second Dynasty. Whatever the purpose and date of the Gisr al-Mudir, however, there can be little doubt that its location reflects the significance of the Wadi Abusir rather than the Nile Valley (fig. 11.6).

Next in the sequence of royal tombs at Saqqara are the Third Dynasty pyramid complexes of Djoser and Sekhemkhet. The locations chosen for both these pyramid complexes also appear to place greater emphasis on the Wadi Abusir than the Nile Valley (fig. 11.6). Although the towering Step Pyramid can be seen from the Nile Valley, this visibility is limited by the landscape, with the limestone escarpment obscuring much of the lower parts of the pyramid. When viewed from the Wadi Abusir, however, the Step Pyramid dominates the landscape (fig. 12.2). When considering the location of the Step Pyramid, it is also important to remember that the original intention was to build a stone mastaba, with only a single tier (chapter 11). Although this lower-lying structure would have been visible from the Wadi Abusir to the west, it would have had little or no visibility when viewed from the Nile Valley in the east.

Sekhemkhet was the last Early Dynastic pharaoh to be buried at Saqqara, and it was not until the Fifth Dynasty that the site was used once more as a royal necropolis. Although by the Fifth Dynasty Egyptian pyramid complexes had adopted the causeway and a generally east–west orientation (fig. 12.3), the Wadi Abusir may still have retained some significance, with the pyramid of Userkaf, the earliest Old Kingdom pyramid to be built at Saqqara, also visible from the wadi (fig. 12.2).

When the sequence of early development at Saqqara is examined, it appears that the Wadi Abusir became increasingly important after the First Dynasty, with all subsequent important structures of the Early Dynastic Period built along its margins. Today the shallow wadi is not a dominant landscape feature, but the vegetation that would have extended along its length in the Early Dynastic Period would have transformed it into a broad swath of green, an extension of the Black Land that offered worshipers at the site a safe route of access, protected from the chaos of the surrounding desert. A few years after I had first published my ideas on the influence of landscape on Early Dynastic development at Saqqara, another paper was published, addressing

Figure 12.2. The Step Pyramid viewed from the Wadi Abusir. The later Old Kingdom pyramid of Userkaf can be seen to the left.

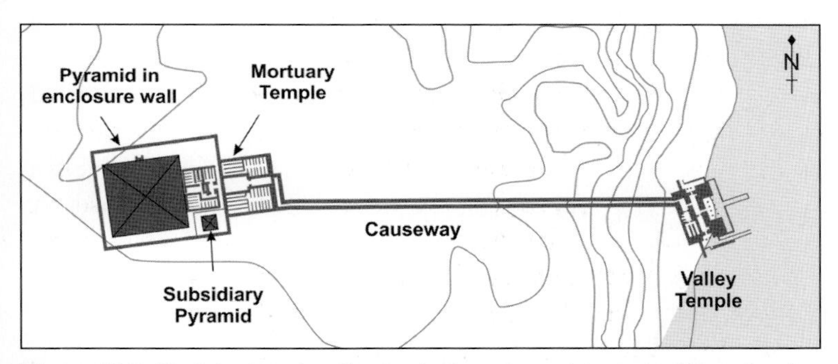

Figure 12.3. Sketch showing the typical east-west layout of the classic Egyptian pyramid complex. Except for the earliest step pyramids, most royal pyramids in Egypt adopted this general layout.

the location of a place referred to by the ancient Egyptians as Ankh Tawy—Life of the Two Lands. Ankh Tawy seems to have been a very significant site for the ancient Egyptians, so much so that it is often thought to be a reference to Memphis, the ancient capital. The location of Ankh Tawy, however, has been the subject of much ongoing debate, with several researchers using their interpretation of ancient texts to associate it with the North Saqqara Plateau. Drawing on my work on the impact of landscape on the development of Saqqara in the Early Dynastic Period, and considering a range of both textual and archaeological evidence, Francisco Borrego Gallardo has recently identified Ankh Tawy as the vegetated Wadi Abusir (see Further Reading), emphasizing the important role that this landscape feature played, not just in the early development of the Saqqara necropolis, but also in the more general theology of ancient Egypt.

My interest in the landscape of Egypt and the role it played in the development of the pharaonic civilization had been inspired initially by a particular element of the "typical" pyramid complex, the causeway. Egypt's earliest pyramids, the step pyramid complexes of Djoser and Sekhemkhet at Saqqara, did not include a causeway; however, from the beginning of the Fourth Dynasty, almost every royal pyramid in Egypt featured a causeway as a core component of the pyramid complex. As shown in figure 12.3, the causeway linked the valley temple, at the edge of the inundation, to the mortuary temple and pyramid, located on the fringes of the desert. The earliest examples of causeways from Meidum and Dahshur (fig. 11.1), however, included a rather puzzling feature that could have served no practical purpose: along these causeways, the ancient builders had covered a durable limestone pavement with a far less robust layer of mud plaster or mudbrick (fig. 12.4).

I see it as no coincidence that the concept of the mud-paved pyramid causeway emerged shortly after the Early Dynastic Period, as Egypt's climate began to dry out and wadi vegetation began to recede. As I have discussed in my paper "On Pyramid Causeways" (see Further Reading), the use of mud plaster or mudbrick paving along the earliest causeways is likely to have replicated the dark soils of the Black Land and was intended to identify the causeway as a route of safe passage through *deshret*, a function that at Early Dynastic Saqqara had been fulfilled by the vegetated Wadi Abusir. As the pyramid complex in ancient Egypt developed further, the symbolism of the darkly paved causeway also developed. The soft mudbrick and mud plaster of the earliest causeways were replaced by an equally dark but far more durable product of Egypt's landscape, black basalt from Widan al-Faras (figs. 10.9 and 11.1).

Figure 12.4. The upper end of the causeway of the Bent Pyramid at Dahshur (see also figure 11.7, left). Although the causeway was dressed with limestone paving, this was overlaid with a far less durable mud or mudbrick paving, surviving elements of which are visible as the dark-colored areas in this photograph.

The darkly paved causeway is just one of many examples of the influence of landscape on the early development of the pharaonic civilization. Although, as we explored in chapter 6, the climate of Egypt may have changed and the River Nile continued to meander within the confines of the Eonile Canyon, the importance of landscape for the theology of ancient Egypt never diminished. Although this book generally addresses only the earliest phases of construction in ancient Egypt, structures such as the New Kingdom temple of Karnak illustrate the fundamental role that geology and landscape continued to play in the architecture and theology of this extraordinary civilization. The immense, closely spaced sandstone columns of Karnak's Great Hypostyle Hall, for example (fig. 12.5), were built to represent a papyrus swamp, a hugely important landscape feature in the theology of ancient Egypt, from which the great god Atum is said to have emerged at the very beginning of creation.

Figure 12.5. The columns of Karnak's Great Hypostyle Hall represent the primeval papyrus swamp in which ancient Egypt was created.

CHAPTER 13
A GIFT OF GEOLOGY

The story told in this book covers inconceivable periods of time. We started thousands of millions of years ago and explored how the landscape of Egypt has evolved since that remote period, how continents have collided and seas have waxed and waned, progressively building up the layers of rock that form the basis of today's remarkable landscape. Briefly, we explored the environment that existed at particular points in time and the animals that inhabited those environments, from dinosaurs to early whales, and the dramatic evidence for these animals that has been found in the deserts and oases of modern Egypt. Then came the upheavals that led to the opening of the Red Sea and, most significantly for Egypt, the formation of the surrounding highland regions including the Red Sea Hills, with their complex geology and vast mineral wealth. We then explored our current understanding of the river systems that drained this evolving landscape, ultimately leading to the development of the River Nile, one of the most significant rivers on the planet today. Finally, we discussed the changing climate over the last 12,000 years and its influences on the people of ancient Egypt.

Figure 13.1. Sunburst over the Nile.

Why is any of this important to Egypt, to its history, its present, and possibly its future, and what was it that set Egypt apart from other river-based civilizations of the time? To answer that question, we have delved into a number of research topics, all of which link the development of the pharaonic culture to aspects of Egypt's landscape. Herodotus' oft-quoted view that "Egypt was a gift of the Nile" represents just one of these narratives, and while it is undeniably true that the abundant fertility of the Nile Valley allowed the people who lived there to make a very good living off the land, these people were not ignorant of the varied landscape that lay beyond the Nile. When the pharaonic culture first emerged from the mists of prehistory in the fourth millennium BC, Egypt's climate was in transition. As conditions became increasingly arid and the desert environments that border the Nile became more hostile, it is likely that cultures from areas that are now desert were drawn toward the Nile and the oases, and they were subsumed into the pharaonic population, adding richness and variety to a complex system of cultural beliefs, one that still is not fully understood by modern scholars.

Although it was not the only determining factor, Herodotus was right to focus on the significance of the Nile for the development of the pharaonic culture. Perhaps the single most influential aspect of the great river is the quirk of geography that caused it to flood once a year, renewing the fertility of the soils and making agriculture in the Nile Valley a seasonal activity. Out of this seasonal activity came a class of learned people who designed a calendar and watched the stars to predict when the next flood would arrive so that the people of ancient Egypt could prepare. Not that every flood was like the last. Abnormal floods could lead to disaster, and so the learned classes began to record the level of the floodwaters for their pharaoh and his ministers. All of this required systems of state administration and the development of writing.

Another vitally important element of the pharaonic culture that was shaped by Egypt's landscape was the emergence of craft specialization. Although the agricultural land of the Nile Valley was submerged for significant parts of the year, the fertile soils that were deposited as the floods receded allowed the people that worked

Figure 13.2. Shafts of sunlight pierce the gloom of the New Kingdom Temple of Seti I at Abydos.

the land to produce more than was required to meet their immediate needs. This meant that they could trade their surplus for other goods, allowing other members of the society to focus on nonagricultural skills and crafts. This also meant the introduction of a system of taxation, which paid for the functioning of the pharaonic state and contributed to ancient Egypt's great wealth. Out of this, an artisan class developed that included potters, scribes, stoneworkers, and metalworkers; these artisans became so accomplished that even today, in the twenty-first century AD, we marvel at what they achieved. How did they build the pyramids? How did they carve beautiful hieroglyphic texts so precisely into stones as hard and resistant as granite? How did they become such accomplished goldsmiths? Many of these achievements cannot be replicated today in our world driven by economics and market forces.

The natural resources these artisans used were all part of Egypt's landscape, a landscape that included a wondrous variety of rock types, from general building stones in abundance along the Nile Valley, to more exotic stones such as the granites of Mons Claudianus deep in the Eastern Desert that were coveted by the people of distant Rome for their most prestigious buildings. Egypt was abundant in metals such as copper and gold, as well as gemstones such as amethyst, turquoise, and carnelian. Mined largely in the Eastern Desert, this natural mineral wealth was embraced by the people of the Nile Valley, who produced wondrous examples of the jewelers' art, such as the famous burial mask of Tutankhamun (fig. 1.7). The bounteous agricultural and mineral wealth that blessed the people of ancient Egypt was undoubtedly the main factor that led to its dominance of the Near East, particularly in the New Kingdom.

There are doubtless other factors that influenced the emergence and development of the pharaonic cultures of Egypt. Whichever of those strands we choose to focus on, however, I suspect that their fundamental underpinnings will lie in the geology and landscape of Egypt. It is for this reason that I conclude that Herodotus was only partly right when he described Egypt as "a gift of the Nile." The origins of ancient Egypt lie not just in the Nile, but in the landscape through which the waters of the Nile flowed. Egypt and the ancient Egyptian civilization are truly "gifts of geology."

FURTHER READING

Print Sources

Arnold, D. *Building in Egypt: Pharaonic Stone Masonry*. Oxford: Oxford University Press, 1991.
Perhaps the most comprehensive work on methods of construction in ancient Egypt.

Baines, J., and J. Málek. *Atlas of Ancient Egypt*. Oxford: Andromeda, 1980.
An authoritative and comprehensive work on the archaeological sites of Egypt.

Bard, K.A. *Encyclopaedia of the Archaeology of Ancient Egypt*. London: Routledge, 1999.
A comprehensive reference work covering many aspects of ancient Egypt.

Bárta, M., and M. Frouz. *Swimmers in the Sand: On the Neolithic Origins of Ancient Egyptian Mythology and Symbolism*. Prague: Dryada, 2010.

An exploration of the rock art of the Western Desert and its possible influences on pharaonic culture.

Borrego Gallardo, F.L. "Ankhtawy: Notes on Its Nature and Location between the Old and New Kingdoms." ISIMU 20-21, 2018.
A research paper that identifies the Wadi Abusir at Saqqara as the location of the ancient site of Ankh Tawy.

Bowen, G.E., and C.A. Hope. *The Oasis Papers 9: A Tribute to Anthony J. Mills for Forty Years of Research in Dakhleh Oasis*. Oxford: Oxbow Books, 2019.
Proceedings of a recent conference summarizing the latest understanding of the archaeology of the Western Desert, with emphasis on the Dakhla and Kharga oases.

Brier, R., and J-P. Houdin. *The Secret of the Great Pyramid*. New York: Smithsonian Books, 2009.
A description of the theory that the Great Pyramid of Khufu may have been built using a system of internal ramps.

Butzer, K.W. *Environment and Archaeology: An Ecological Approach to Prehistory*. Chicago: Aldine-Atherton, 1979.
An academic assessment of the influences of changing global ecology on our understanding of archaeology.

Clarke, R. *Tomb Security in Ancient Egypt from the Predynastic to the Pyramid Age*. Oxford: Archaeopress Archaeology, 2016.
A study of the measures adopted to protect tombs in the early part of the pharaonic era.

Clayton, P.A. *Chronicle of the Pharaohs*. London: BCA, 1994.
A comprehensive account of the lives and accomplishments of the known pharaohs of ancient Egypt.

Cross, S. "The Hydrology of the Valley of the Kings." *The Journal of Egyptian Archaeology* 94, no. 1 (2017): 303–10.

A technical paper presenting an interpretation of the geology and hydrology of the Valley of the Kings and its implications for the survival of Tutankhamun's tomb.

Davies, V., and R. Friedman. *Egypt*. London: British Museum Press, 1998.
A discussion of significant discoveries and archaeological fieldwork in Egypt.

Der Manuelian, P. "On the Early Lost History of Giza: The 'Lost' Wadi Cemetery." *Journal of Egyptian Archaeology* 95, no. 1 (2009): 105–40.
A technical paper exploring evidence for a series of tombs that predate Khufu's Western Cemetery at Giza.

Dodson, A. *Ancient Egypt*. London: New Holland, 2006.
A comprehensive account of the pharaonic culture.

Edwards, I.E.S. *The Pyramids of Egypt*. London: Penguin, 1993.
The classic work on the pyramids of Egypt.

Fakhry, A. *The Pyramids*. Chicago: University of Chicago Press, 1961.
An extremely important work on Egyptian pyramids.

Forman, W., and S. Quirke. *Hieroglyphs and the Afterlife*. London: British Museum Press, 1996.
A lavishly illustrated work celebrating some of the highlights of pharaonic culture.

Förster, H., and F. Riemer. "Ancient Desert Roads: Towards Establishing a New Field of Archaeological Research." In *Desert Road Archaeology*, edited by H. Förster and F. Riemer, 19–58. Africa Praehistorica 27. Cologne: Heinrich-Barth-Institut, 2013.
A collection of technical papers summarizing the latest understanding of desert road archaeology in Egypt and further afield.

Friedman, R., ed. *Egypt and Nubia: Gifts of the Desert*. London: British Museum Press, 2002.
A collection of technical papers exploring the archaeology of Egypt's desert regions presented at a colloquium held at the British Museum in 1998.

Hawass, Z., ed. *Egypt at the Dawn of the Twenty-first Century*. Cairo: American University in Cairo Press, 2003.
Technical proceedings of a conference held at the turn of the century to address the latest thinking on a range of Egyptological matters.

Heath, J.M. *Before the Pharaohs*. Barnsley: Pen and Sword Books, 2021.
An up-to-date account of the archaeology of Stone Age Egypt.

Hillier, J.K., et al. "Monuments on a Migrating Nile." *Journal of Archaeological Science* 34, no. 7 (2007): 1011–15.
A technical paper presenting an account of the investigations of the movement of the Nile channel during and after the pharaonic era.

Huyge, D., et al. "Dating Egypt's Oldest 'Art': AMS ^{14}C Age Determinations of Rock Varnishes Covering Petroglyphs at El-Hosh (Upper Egypt)." *Antiquity* 75, no. 287 (2001): 68–72.
A technical paper presenting an innovative technique for dating rock art.

Kemp, B. *Ancient Egypt: Anatomy of a Civilisation*. London: Routledge, 1991.
Perhaps the most masterly account of the development and features of the pharaonic culture.

Klemm, D., and R. Klemm. *The Stones of the Pyramids: Provenance of the Building Stones of the Old Kingdom Pyramids of Egypt*. Berlin and New York: De Gruyter, 2010.

A technical publication presenting evidence for the source of the building stone used in Egypt's major pyramids.

Kramers, J.D., et al. "Unique Chemistry of a Diamond-bearing Pebble from the Libyan Desert Glass Strewnfield, SW Egypt: Evidence for a Shocked Comet Fragment." *Earth and Planetary Science Letters* 382 (2013): 21–31.
Technical paper exploring possible methods of formation of Libyan Desert glass.

Lauer, J.-P. *Saqqara*. London: Thames and Hudson, 1978.
A detailed overview of the Saqqara necropolis written by the man who dedicated most of his life to its investigation and restoration.

Lehner, M. *The Complete Pyramids*. London: Thames and Hudson, 1997.
A comprehensive and readily accessible review of the pyramids of Egypt.

Maxfield, V.A., and D.P.S. Peacock, eds. "Survey and Excavation: Mons Claudianus 1987–1993. Volume 3, Ceramic Vessels and Related Objects." *Fouilles de l'Institut Français d'Archéologie Orientale* 54 (2006): i–450.
A comprehensive academic work on excavations and other studies undertaken at this Roman quarry in the Eastern Desert.

Maxwell, E., et al. "Evidence for Pleistocene Lakes in the Tushka Region, South Egypt." *Geology* 38, no. 12 (2010): 1135–38.
A technical paper exploring former aspects of surface hydrology in the Western Desert.

Nicholson, P.T., and I. Shaw, eds. *Ancient Egyptian Materials and Technology*. Cambridge: Cambridge University Press, 2000.
A comprehensive academic work on the materials and technology used in ancient Egypt.

Nothdurft, W., and J. Smith. *The Lost Dinosaurs of Egypt*. New York: Random House, 2002.
A thrilling account of the discovery and rediscovery of dinosaur fossils in Bahariya Oasis.

O'Connor, D. *Abydos*. London: Thames and Hudson, 2009.
A detailed overview of this important ancient site written by one of the foremost archaeologists working there.

Reader, C.D. "An Early Dynastic Ritual Landscape in North Saqqara: An Inheritance from Abydos?" *Journal of Egyptian Archaeology* 103 (2017): 71–87.
A technical paper discussing the influence of landscape on the Early Dynastic development of the Saqqara necropolis.

Reader, C.D. "A Geomorphological Study of the Giza Necropolis, with Implications for the Development of the Site." *Archaeometry* 43, no. 1 (2001): 149–65.
A technical paper presenting geological evidence to indicate that the Great Sphinx is older than the Giza pyramids.

Reader, C.D. "The Meidum Pyramid." *Journal of the American Research Center in Egypt* 51 (2015): 203–24.
A technical paper discussing the evidence that the Meidum Pyramid was initially built by the Third Dynasty pharaoh Huni.

Reader, C.D. "On Pyramid Causeways." *Journal of Egyptian Archaeology* 90 (2004): 63–71.
A technical paper discussing the origins and purpose of the causeway as part of the ancient Egyptian pyramid complex.

Reeves, N. "The Burial of Nefertiti?" *Amarna Royal Tombs Project, Valley of the Kings*. Occasional Paper 1 (2015): 1–16.
A technical paper setting out the evidence for possible further elements hidden in Tutankhamun's tomb in the Valley of the Kings.

Rice, M. *Egypt's Making*. London: Routledge, 2003.
A wonderful account of the formative years of the pharaonic culture.

Riemer, H., et al. "Climate, Styles and Archaeology: An Integral Approach Towards an Absolute Chronology of the Rock Art in the Libyan Desert (Eastern Sahara)." *Antiquity* 91, no. 355 (2017): 7–23.
A technical paper presenting evidence for the age of the various rock art cultures of Egypt's Western Desert.

Rothe, R.D., et al. "New Hieroglyphic Evidence for Pharaonic Activity in the Eastern Desert of Egypt." *Journal of the American Research Center in Egypt* 33 (1996): 77–104.
A technical paper discussing pharaonic-era inscriptions in the Eastern Desert and the evidence this may provide for the exploitation of specific mineral resources in the area.

Said, R. *The Geological Evolution of the River Nile*. New York: Springer-Verlag, 1981.
An academic work on the evolution of the River Nile.

Sampsell, B. *The Geology of Egypt*. Cairo: American University in Cairo Press, 2014.
An excellent work providing an introduction and overview of Egypt's geology.

Saul, J.M. "Flickering Flames over the Libyan Desert?" *International Geology Review* 61, no. 11 (2018): 1340–69.
A recent technical paper providing a comprehensive review of the theories of formation of Libyan Desert glass.

Schulz, R., and M. Seidel. *Egypt: World of the Pharaohs*. Cologne: Könemann, 2010.
An encyclopedic review of our current understanding of pharaonic Egypt, written by some of the subject's leading scholars.

Shaw, I., ed. *The Oxford History of Ancient Egypt*. Oxford: Oxford University Press, 2000.
An authoritative review of the history of ancient Egypt.

Stanley, J.-D., et al. "Nile Flow Failure at the End of the Old Kingdom, Egypt: Strontium Isotopic and Petrologic Evidence." *Geoarchaeology* 18, no. 3 (2003): 395–402.
An academic work on the evidence for low Nile floods influencing the demise of the Old Kingdom.

Stocks, D. *Stoneworking Technology in Ancient Egypt*. London: Routledge, 2003.
A technical account of research undertaken into the use of copper in stoneworking and other tools in ancient Egypt.

Sullivan, E.A. "Seeking a Better View: Using 3D to Investigate Visibility in Historic Landscapes." *Journal of Archaeological Method Theory* 24 (2017): 1227–55.
A technical paper exploring the landscape of Memphis and what 3D computer modeling may tell us about human perception of monuments in the landscape.

Tallet, P., and M. Lehner. *The Red Sea Scrolls*. Thames and Hudson, 2022.
A popular account of the discovery of Old Kingdom papyri at Wadi al-Jarf and the implications for our understanding of the development of the Giza necropolis.

Tallet, P., and G. Marouard. "The Harbour of Khufu on the Red Sea Coast at Wadi al-Jarf, Egypt." *Near Eastern Archaeology* 77, no. 1 (2014): 4–14.
A technical paper summarizing a series of remarkable recent finds on the shores of the Red Sea.

Weeks, K., ed. *Valley of the Kings*. Vercelli: White Star, 2001.
A lavishly illustrated account of the Valley of the Kings, written by one of the leading authorities on the site.

Wengrow, D. *The Archaeology of Early Egypt*. Cambridge: Cambridge University Press, 2006.
A comprehensive academic account of the formative years of the pharaonic culture.

Wilkinson, T.A.H. *Early Dynastic Egypt*. London: Routledge, 1999.
Possibly the most comprehensive account of Early Dynastic Egypt.

Wilkinson, T.A.H. *Genesis of the Pharaohs*. London: Thames and Hudson, 2003.
An authoritatively written popular account of the origins of the pharaonic civilization.

Online Resources

The Amarna Project. https://www.amarnaproject.com/

Harrell, J. "Quarry Scapes: Conservation of Ancient Stone Quarry Landscapes in Eastern Mediterranean." http://www.quarryscapes.no/top_links.php

Hierakonpolis Online. https://www.hierakonpolis-online.org/

Torcia, M. "Giza and Hierakonpolis: *Cretulae* with Figurative Seal Impressions and Isolated Signs of Writing. Connections with the Mesopotamian Area." https://www.academia.edu/25349666/Giza_and_Hierakonpolis_cretulae_with_figurative_seal_impressions_and_isolated_signs_of_writing_Connections_with_the_Mesopotamian_area.
A technical paper that examines the age of a number of clay seals found during excavations at Giza.

IMAGE CREDITS

1.6.	Dates taken from J. Baines and J. Málek, *Atlas of Ancient Egypt* (Oxford: Andromeda, 1980), 36-37.
1.7.	© Reg Clark, used with permission.
2.3.	Based on https://jabradley101.weebly.com/sedimentary.html.
2.4a.	Based on https://www.slideserve.com/tekla/sea-floor-spreading, slide 9.
2.4b.	Based on M. Marot et al., "An Intermediate Depth Tensional Earthquake (Mw 5.7) and Its Aftershocks within the Nazca Slab, Central Chile: A Reactivated Outer Rise Fault?," *Earth and Planetary Science Letters* (2002), figure 4.
3.1, 4.1, 4.2.	Based on global reconstructions provided in H. Schandelmeier and P.-O. Reynolds, *Palaeogeographic-Palaeotectonic Atlas of North-Eastern Africa, Arabi, and Adjacent Areas* (Rotterdam: A.A. Balkema, 1997).

10.1. Based on J. Baines and J. Málek, *Atlas of Ancient Egypt* (Oxford: Andromeda, 1980), 21, and Egyptian Geological Survey and Mining Authority, *Metallogenic Map, Arab Republic of Egypt* (1998).

11.2. © Mike Shepherd, used with permission.

11.3. Dates taken from J. Baines and J. Málek, *Atlas of Ancient Egypt* (Oxford: Andromeda, 1980), 36-37.

Plate 3. Based on the Egyptian Geological Survey and Mining Authority, *Geological Map of Egypt* (1981). Accessed from https://esdac.jrc.ec.europa.eu/content/geologic-map-egypt and reproduced under the Creative Commons Attribution 4.0 International license.

Plate 12. The reconstruction of Lake Tushka is based on E. Maxwell et al., "Evidence for Pleistocene Lakes in the Tushka Region, South Egypt," *Geology* 38, no. 12 (2010): 1135–38.

Plate 18. © Robert Partridge, used with permission.

All other images: © the author.

INDEX

Illustrations in **bold**